From Our Own Correspondent

From Our Own Correspondent

Dispatches of a Decade From Across the World

Edited by Polly Hope

WEIDENFELD & NICOLSON

First published in Great Britain in 2020
by Weidenfeld & Nicolson

10 9 8 7 6 5 4 3 2 1

By arrangement with the BBC

The BBC logo and Radio 4 logo are registered trade marks of the British
Broadcasting Corporation and are used under licence.

A CIP catalogue record for this book
is available from the British Library.

HB ISBN   978 1 4746 0766 7
eBook 978 1 4746 0767 4

Typeset by Input Data Services Ltd, Somerset

Printed in Great Britain by Clays Ltd, Elcograf S.p.A.

Weidenfeld & Nicolson

The Orion Publishing Group Ltd
Carmelite House
50 Victoria Embankment
London, EC4Y 0DZ

An Hachette UK Company

www.weidenfeldandnicolson.co.uk
www.orionbooks.co.uk

Contents

Contents

PART II: AFRICA

Africa 63

Contents

Contents

Contents

Contents

PART VII: THE WHOLE PLANET

The Whole Planet 353

Introduction

Over more than sixty years on the air, *From Our Own Correspondent* has done its best to take listeners to parts of the world where they had never been, and perhaps never would go in their lives: war zones, refugee camps, elite universities, space stations, spy academies and lions' dens of all sorts. Its dispatches introduce audiences to people they might never expect to meet – kingpins, revolutionaries, assassins and outcasts. It has always relied on the power of personal testimony, with its contributors not merely reporting the news, but sharing what they found out along the way, and how it felt. Very often, it was their individual impressions and responses which made a piece vivid and memorable; the programme prides itself on relaying 'the stories behind the headlines' – that is to say, the deeper and longer currents which throw up the crashing waves of current events. 'We need the sights, sounds and smells, as well as the facts and figures': that's how the production team often sums it up for potential writers.

This unusual amount of leeway allowed for personal reactions does set it apart from many other news and current affairs programmes, on the BBC or other broadcasters. Correspondents for *FOOC* are positively encouraged to break the usual mould of news reporting (though some might describe it as being more like bursting free of a corset, or even a straitjacket). Over the decades, the usual trappings of news reporters at work have become predictable, even ritualised: there they will be, standing neatly dressed outside a landmark building, next to a satellite dish, with microphone in hand, talking confidently in assertive 'outside' voices, announcing the headlines or the latest updates to the world. When other voices break in, or the unexpected happens, the spell is broken, and the old formats can begin to look a little superannuated. But try as they might to remake the news in a more fluid, interactive fashion, indulging in more back-and-forth with their audiences, a wider pool of voices, and a sense of constant movement, news broadcasters can't completely jettison the established ways of doing things.

The usual style of a dispatch for *FOOC* may seem even more traditional, verging on outdated. Five whole minutes of a single voice, reading aloud from a script, and nothing else? There aren't many places left in any medium for such an extended, focused exercise. But the very simplicity of the format allows for something different, and often far more intimate than other kinds of reporting; not by attempting to sum up the whole picture, but rather, trying to find another window on it, often by focusing on the smaller details which serve as microcosms of an entire, sometimes terrifyingly large narrative. The Egyptian sofa overlooking Tahrir Square that's seen the entire Arab Spring and its aftermath unfold. The frozen basement full of rapidly thawing whale meat revealing how fast Alaska's permafrost is melting. The empty lunch tables which betray the impact of the financial crisis in Portugal.

It's a programme which demands an unusually high level of literary feeling, and of attention to detail. Those who think of radio as a succession of spontaneous but revealing live conversations would wonder at the extraordinary pains taken on *FOOC* to make each image tell, ensure each description conjures a clear picture, choose precisely the right quote. And then come the fact-checking, the pulling-apart of statistics and the wrangling over the finer points of vocabulary, syntax and grammar. To meld all this together, while still sounding as though you're speaking your own thoughts in your own words, remains a delicate art.

Once BBC foreign correspondents were nearly all of a recognisable type – middle-aged, middle-class, highly literate men from Great Britain, who'd often moved into broadcasting after careers in print journalism, and could be relied upon to interpret strange goings-on abroad for a domestic audience. As British society and the BBC itself changed, since the 1960s, there's been a steady increase of 'other' voices – women, younger people, non-Britons, British citizens with international backgrounds – among their ranks.

In the past few decades, the BBC, like many news organisations, has moved towards using far more 'insiders' – reporters born and raised in the countries they're reporting on, or who've lived there for years, or who are at least specialists in their politics or economy. Or they may come from the very communities they're covering. That background can certainly provide windows which visiting outsiders cannot. There's a vivid, lived-in quality to some reports in this anthology which no outsider could summon, however well informed.

The insights and angles these correspondents have shared would have been difficult for the older model of a 'classic BBC man' to glean. Secunder Kermani's reflection on the kind of haircut he favours – along with seemingly every other man under thirty in one Pakistani town – can tell you more about migration, marriage and aspiration between the UK and South Asia than any number of hard news reports from Westminster or Islamabad. The Western press might editorialise at length about how Russian law and society treat gay citizens, but it's something else again to read Paul Henley's account of shaking the hands of people who (in their own words) would never accept his sexuality.

After a great deal of soul-searching about the ethics of star correspondents 'flying in to do a disaster', there's a real and increasing appetite for international affairs coverage by people who know a story inside out. How can someone based in a news bureau thousands of miles away, who has jetted in for a few days of intensive coverage and then leaves straight afterwards, really have special insight to offer? Why should it be their voice, rather than anyone else's, that we listen to?

This is always a difficult editorial tightrope to walk. While an 'insider' might have the deepest understanding of the story, they might not know the most elegant way to tell it. An 'outsider' just flying in on behalf of the BBC may sometimes get an easier ride from interviewees, or have a wider, more impartial view of a knotty issue. And while there will always be times when only an insider's knowledge will do, there will also be occasions when a British audience really needs someone to express the local nuances of a foreign story in terms it can grasp. There will never be a one-size-fits-all solution.

And what about the audiences beyond the UK – both for the BBC as a whole and for the programme? *From Our Own Correspondent* is now broadcast on the World Service, too, in an international edition, which may frame the stories differently but uses many of the same dispatches. Finding the right words, the right level of detail, the right register to tell a story as well as possible, to as many people as possible, will always be an exercise in compromise and consideration.

Notoriously, a great blizzard of names and numbers is never likely to work well in audio; hard data is often easier to grasp in visual form, while a complex investigative paper trail might be easier to follow in written prose than through an exhaustive spoken account of every step. But what the kind of radio reporting which *FOOC* specialises in really can achieve – perhaps more vividly than any other medium – is connecting listener and journalist with extraordinary intimacy. In retelling their own account of how they got a news story, reporters can zero in on what really mattered: the images and words they'll never forget, and the moments which made the deepest impression on them. That personal touch is what makes the finest dispatches on *FOOC* really sing.

Witnessing history, weighing up what really matters and then writing about it as it's still happening all demand stamina, as well as lightness of touch. Most journalists are quick to recognise their own privilege in being able to choose whether or not to go into a war zone, report on a natural disaster or cover a human emergency – and, more crucially, in being able to leave. This demands toughness. Their job is to get and relay the facts, not to act as relief workers, and certainly not to be self-indulgent about their own feelings. Yet you can't report with accuracy if you have no empathy.

In August 2015, after weeks on the so-called 'migrant trail' through south-eastern Europe, covering the largest mass exodus there since the end of the Second World War, Nick Thorpe considered what he'd seen, and what he was really doing: 'One of the most exhausting aspects of this refugee drama for us reporters,' he wrote, 'is that we make ourselves available as story tunnels. We funnel the plight of individual men, women and children into Europe, ahead of them – and we funnel back the fears and concerns of Europeans to the refugees' countries of origin, and to the sprawling camps surrounding them. We repeat and interpret the words of the politicians who want them in, or want them out, or can't decide. For two months I haven't had a non-migrant dream . . .' His words apply to many others who are still out there 'in the field', working in conditions of extreme stress and sometimes physical danger.

In sifting through a decade's worth of dispatches for this book, it soon became clear that simply pulling together snap responses to 'headline events' wouldn't do. As *FOOC*'s producers often point out, 'life's not all wars and elections, you know!' Some correspondents really had been in exactly the right place at the right time, and witnessed the 'big events' or 'defining moments' at first hand. More often, they would arrive as soon as they were able and do the best they could to get as close as they could to the story. Often the most revealing writing came about not in the heat of the moment – when they'd be busy juggling hourly pieces to camera, relentless calling and texting, and the demands of their own technology – but in the quiet hours of the morning, on the flight home, or perhaps months or weeks (even years) later, after mulling it all over. (As an aside: it may seem odd that Brexit plays no part in this compilation, but as *FOOC* is a programme of international current affairs, and Brexit is a contentious matter of UK government policy, it's a subject the radio broadcasts have largely avoided.)

So what have the correspondents been covering over the past ten years? It soon became obvious that four major, intricately interlocked 'big stories' stood out: the Arab Spring and its aftermath, the financial crisis, the war in Syria, and the surge in migration to Europe during 2014–2015. Each of those crises generated avalanches of news coverage, and correspondingly urgent dispatches on *FOOC*. But then there were the other, even bigger stories shaping our world: more amorphous and slow-moving, but even larger-scale narratives which would surely affect far more lives in their time. China's rise to strategic and economic superpower status. Climate change and its consequences. The new landscapes of the earth as human consumption and settlement threatened entire cultures, species and ecosystems. And a relentless background hum, heard in story after story: the way technology was remaking everyone's social and political life.

One telling – but internal – indicator reveals this shift. As part of the inevitable paperwork prepared for every episode of *FOOC* to be broadcast, the producer must list any mentions of brand names or commercial entities made in each programme. From the 1980s to the early 2000s, these were usually indicators of brute economic or military power; whenever there was a civil war on anywhere, references to Mercedes-Benz cars and Kalashnikov rifles would crop up in almost every episode. But by 2010 the world had grown noticeably more digital, and the most frequently cited brands had become the virtual titans of the information economy: iPhone, Facebook, Twitter, Google, YouTube.

That transformation is going to change events themselves, as well as the way they are reported, in the decades ahead. 'User-generated content' (the blurry videos shot on smartphones, the screenshots of revealing instant messages), rather than old-fashioned journalists' shoe-leather or fixed phone lines, is often now the most direct route into a major story. Audiences now have a level of direct access to others' lives unthinkable during the early years of radio or television broadcasting. If knowledge really is power, then the balance of power between broadcasters and their audience is clearly shifting, and so is the relationship between those audiences and the people who govern them. All that makes the dangers of 'fake news' ever more pressing, as partisan, incorrect or deliberately skewed report- ing can bring about real-world results in record time. Mass panics, communal riots and sharpened prejudices have all been brought on by viral news.

A dispatch from Yemen encapsulated this. In February 2017, Elizabeth Kendall wrote of her travels in a remote region with some local bodyguards: 'On my first visit to these deserts, almost five years earlier, they had spoken about Yemen itself as though it were a distant foreign country. Now they were asking me what I thought about the oppression of Rohingya Muslims in Myanmar and the anti-Muslim campaign promises of Donald Trump. What had changed, I asked? *"Faysbuuk . . . WhatsAbb!"* they laughed, as they whipped out their new smartphones.'

It's undeniable that technology has made radio journalism easier in some ways – not least in reducing the weight of a reporter's kit. In the old days they'd have to haul around reel-to-reel recorders the size and weight of Victorian suitcases; today we can take in dispatches from journalists who've recorded them using their mobile phones, and then sent the audio halfway around the world as an email attachment. Yet the endless multiplication of different platforms – radio, television, the BBC's own news website, podcasts, social media – also makes far more demands on correspondents' time. The competition is fierce and the pace is blistering. But there's always still a place for one of the oldest human appetites: the hunger for a good story, told by someone who knows how.

So this book shouldn't be read simply as a comprehensive digest, much less a ranking of the biggest news events of the decade, nor as a connoisseur's selection of the finest writers to contribute to *From Our Own Correspondent*. Instead, it's more like a compilation of personal snapshots: of those telling moments and revealing details which throw an intimate light on how the world is changing. Each edition of the programme on air tends to meander around the world following its own logic; the only real rule is not to be boring. Thus this book, while roughly divided into geographical sections, takes the same approach: it begins in Europe and travels roughly eastward around the planet, in no particular order, before drawing its breath to consider some stories on a truly global scale.

Part I:
EUROPE

European Union

Chris Bockman – France – Remaking the Beret – 18/05/11

One of the best ways to tell a huge, global story in five minutes on the radio is to home in on a particular object which carries a weight of cultural expectation. From the hijab to an admiral's tricorn, headwear expresses a great deal about who we think we are and how we'd like to be seen. The beret, once a bit of shorthand for French identity, provides an object lesson in how once-treasured symbols of European culture have changed their meaning – and had to reinvent themselves. Chris Bockman went to its spiritual home . . .

Although the beret is one of France's most totemic objects the world over, its historical roots lie in the small town of Nay, deep in the Pyrenees mountain range. Its wide streets, arcades and large warehouses give away that this was once an important manufacturing centre. After the Second World War, the town and surrounding region was home to fifty factories and thousands of jobs in the beret-making business. Now there are just two beret workers left in Nay. They run the local Beret Museum in one of the former grand hat-making factories.

The museum is dedicated to the history of the hat. There are plenty of old looms, once used to make hats from the wool of local sheep. Faded photos on the walls underline that berets were once almost compulsory for men, a proud display of regional identity. Children wore them to school. For farmworkers they provided essential protection against the damp and cold of winter and intense sunshine in the mountains.

The tide began to turn against the beret in the 1950s. Farm labour started to die out, young people began to flock to the cities, and hats in general began to go out of style. But it was globalisation which finished off the French beret for good as a mass-produced export. Like countless other sectors of the European textile industry, production moved to Asia. Nearly all the world's berets are now made in Bangladesh. French beret manufacturers – predominantly based in south-west France – were unprepared for the competition and unable to compete on price. Most of their factories closed down for good in the 1980s.

Just two have survived: Beatex and Blancq-Olibet. Between them they have around a hundred employees. Their headquarters are less than thirty miles apart – but rather than join forces, these two firms watch each other warily and try to keep hold of whatever market share they can.

Still, there may be strategies to keep the French beret industry going. Alain Zachar runs Blancq-Olibet. He used to be in the shirt-making business – but saw that industry vanish too, as low-cost producers emerged in China. So he has gone into berets. But he told me that this time he focused on big buyers with financial backing.

In other words: the military. The French army is buying berets in the tens of thousands and if Alain can provide high-quality, well-made hats tough enough for soldiers, then he could yet win out. And not just in France. He sees potential for selling military berets across the world – especially for UN-peacekeeping operations, where a floppy hat appears far less aggressive than a helmet.

Down the road in Oloron St Marie, the Beatex company is housed in an old factory now far too big for its current production. Modern machines spin wool into yarn. Next door around a dozen women work behind sewing machines, threading together the trimmings.

In a showroom full of the latest models, you quickly see that this company has a new customer in mind. With prices as high as 70 euros per beret, Beatex is targeting women looking for a 'peasant chic' look. Celebrities like Madonna and Claudia Schiffer have been spotted wearing the hat. Luxury designers Christian Dior and Hermès like the idea and have added these high-quality,

French-made berets to their collections, hoping women not just in Paris but in Tokyo and New York will buy them.

Beatex's owner Pierre Lemoine told me that going upmarket is the only way he can compete with cheap imports. The French-made beret only has a future as an expensive fashion accessory, he says. It's certainly come a long way from its traditional roots. How ironic that a garment so rooted in a regional, rural, peasant lifestyle could survive thanks to a global market of urban fashionistas.

Alan Johnston – Italy – Five Star – 01/09/12

It can be hard to spot the end – or the beginning – of a historical era when you're in the thick of it. The fourth of October 2009 isn't a date engraved on many memories, yet for Italians it may prove to have been a watershed. That was the day that long-time provocateur and comedian Beppe Grillo announced the formation of the Five Star Movement (Movimento Cinque Stelle, or M5S) as a political campaign hoping to break what he called the stagnant, post-Second World War consensus in Italian politics. M5S wanted no more of the frequent yet pointless elections which returned the same old faces to power, without making any real changes. It called for no more cronyism, backroom deals or corrupt pacts. Direct democracy was its watchword, and online organising allowed its ideas to spread fast. Its populist appeal was obvious – and within a decade it would be part of a coalition government in Rome.

The mayor's office in Parma's medieval town hall is everything you might expect: a centuries-old seat of power in a wealthy Italian city. You can almost feel the weight of tradition. Aristocrats, dukes and duchesses stare out from old oil paintings. A thick marble desktop balances on the heads of carved golden lions. And a chandelier hangs beneath a vaulted ceiling.

But there's nothing at all traditional about the new mayor who sits beneath the chandelier. Federico Pizzarotti represents the coming of people power in the politics of Parma. A few months ago he was just another commuter: a rather slight, smiling man on his way to a job in IT. But he stood for election as the Five Star Movement's candidate ... and surfed into office on a wave of contempt for the city's regular politicians.

Mr Pizzarotti's never been elected to anything before. But suddenly he finds himself in charge of a quite major city, which is nearly a billion dollars in debt. He acknowledges that of course he and his fellow Five Star councillors have absolutely no experience in this sort of area. But they do have experience of life, he says, and good judgement, and they will simply apply those things to the running of Parma. They know what's in the people's interest.

And that, Mr Pizzarotti would say, is what marks his Movement out from Italy's traditional parties. In the eyes of the Five Star citizens, the political establishment is utterly self-seeking – irredeemably corrupt. A tall, thin, twenty-five-year-old student called Marco Bosi, a spokesman for the Movement on Parma's council, put it like this: 'The party system had failed the city,' he said, 'and the people have taken charge.' All across the country, he went on, too many people didn't feel represented. The solution, he said, was either to emigrate, or to try to change Italy ...

The Movement aims to embody change. It wants to be everything that the established parties are not. There's no Five Star membership list, no offices, no structure, no formally designated leadership. The aim is for the view of every single activist to count as much as any other. And at the Movement's core is the idea that citizens need to participate continually in decision-making. If a hospital is to be built – or to be closed – everyone who might use it has a right to have a say.

But you might argue that faith in the wisdom of crowds can be overdone – that the masses can have moments of madness too. And I asked the Five Star student, Mr Bosi, if there wasn't a danger that in all that mass consultation there'll be masses of talk, but not so much action? That big, controversial decisions might be slow to emerge, that something like paralysis might lie in store for Parma, rather than swift and efficient government? He said that the new regime might be slower to arrive at decisions. But that when it did they would be better, because they'd been endorsed by everyone.

Although the Five Star Movement rejects the notion of hierarchy and leadership, it certainly feels like it has a leader – or at least a sort of 'citizen number one'. He is Beppe Grillo, one of Italy's best-known comedians – a shaggy-haired bear of a man, a bouncing ball of energy, always barracking and blasting the establishment. For example, Mr Grillo advocates a public trial for everybody who's ever been elected to office. Each of them would be made to account for any unexplained wealth – any extra apartments, or the odd, mysterious half-million euros in the wife's bank account.

It's around a stream of thoughts like that that the Five Star Movement has coalesced on the internet, and rapidly become a serious force in Italian politics. And as Beppe Grillo rocks the establishment, he is coming under increasing scrutiny. Some see him as a dangerous populist. One critic has described him as whipping up a wave of feeling that threatens to sweep away the structures of Italian democracy … a raucous Pied Piper leading people to disaster.

Parma is where the Five Star Movement's ideas will be tested for the first time. The new mayor Mr Pizzarotti knows that all of Italy is watching. But he doesn't seem too fazed. He says he'll set about his big challenge in the same way that he would set about trying to eat an elephant. He will do it bit by bit.

Niall O'Gallagher – Spain – Catalan Criticisms – 20/10/12

Over the past decade many an EU member state has wrestled with problems of devolution – the question of where, and at which level, real decisions should be made. Ideals of direct or popular democracy often clashed with the habits of central governments as regions with their own deeply held identies agitated for more powers. Sometimes the demands of cultural or linguistic minorities became more urgent. In Spain, since the end of the Franco dictatorship, there has been an ever-stronger sense of near-autonomy in the 'other' regions. The Basque Country, Galicia, Andalusia and Valencia all squeezed concessions from the national government based in Madrid. But the Catalan sense of regional pride was particularly strong – and emotive. Niall O'Gallagher may be the only BBC employee to speak both Gaelic and Catalan – so he was well-placed to explore the ambivalence and complexity of Catalonian aspirations.

In Barcelona, people have no problem talking to strangers. But then, there can't be many better ways of drawing attention to yourself then carrying a camera and tripod around the city's underground railway system. And, if like me, you have the kind of Celtic complexion that goes red under a sixty-watt bulb and have made the mistake of packing only dark suits to wear in the unseasonable autumn heat, people are curious to know where you've come from and what you're doing there.

I've been a regular visitor and can get by in Catalan – something most locals regard as unusual for a foreigner, but usually approve of, making them more willing to open up. I had one such conversation with a middle-aged couple on the yellow line from the broad boulevards of the city's nineteenth-century Eixample district to the port at the old fishing village of La Barceloneta. It began with the backslapping and jokes about bad weather that seem to accompany Scottish travellers anywhere in southern Europe.

Then we got on to the reason for our visit. 'I'm here', I told them, 'to report on the Catalan independence debate.'

At that point the atmosphere changed. There was no hostility – our couple didn't seem to resent foreigners poking their noses into their business – but while the husband had been taciturn, though smiling, from the start, his warm and elegantly dressed wife, so gregarious just a moment ago, suddenly seemed to have trouble finding her tongue. 'It's difficult,' she told me. 'It's very difficult.'

That was something I came across again and again among the ordinary Catalans I spoke to, particularly those over fifty. They're old enough to remember a time when it wasn't possible to speak freely here, particularly not in Catalan, and certainly not about an issue like this. In the Santa Caterina food market, I put the question to two ladies in their seventies as they haggled over the price and size of the shining cuts of fish being chopped and wrapped before them. 'What do you think, *senyores*, about independence for Catalonia?' One laughed, while the other turned to me, smiled and drew her two fingers across her lips, zipping them shut.

That's a neat image, I think, for the legacy of the dictatorship in Catalonia. Memories of the Franco period hold many people back from speaking openly about independence from Spain. But the written law also stands in the way of any consultation with the people as a whole. The post-Franco constitution calls on the armed forces to defend the unity of Spain. Whether or not Madrid would really be willing to send in the tanks to stop a referendum, the worry that they might is never far from the surface.

There is one place where no rule or reticence inhibits the overt display of pro-independence sentiment. We squeezed onto another crowded metro train to the home of FC Barcelona for their first league clash of the season against Real Madrid. Just weeks after demonstrations which saw an estimated 1.5 million people take to the streets to demand independence from Spain, campaigners used the classic derby to give their arguments a global audience. At exactly fourteen minutes past five that afternoon, supporters inside the stadium raised red and yellow tiles creating a mosaic of the Catalan flag, La Senyera – commemorating the fall of Barcelona in 1714, the year to which Catalan nationalists date the loss of their independence.

'We pay a lot of taxes to Madrid and we don't get enough back,' said one young mother as she pushed her twins towards the gates in a Senyera-decked pram. 'If we had control of our own money, we could give our children a better education.'

'How come the king doesn't speak Catalan if we're supposed to be part of Spain?' asked a lad in a Barca football shirt. 'We've tried to be part of Spain but they don't want us. Now we need independence.'

In the midst of the economic crisis, the Spanish government seems in no mood to negotiate a compromise that might lessen demands for independence on the Catalan streets. And so the scene is set for a confrontation, with no clear sense of when travellers on the Barcelona metro will get to have their say.

Nick Thorpe – Hungary – Orban's National Theatre – 12/04/14

The Viktator. The Orbanator. Hungary's president, elected in 2003 after eight years spent leading his country's right-wing opposition party, Fidesz, has become emblematic of a certain kind of 'post-liberal' leader: a populist strongman, firmer on questions of national pride than free-market economics. An anti-globalist, proud to speak out in favour of what he calls 'illiberal democracy' and to resist outside pressures on his administration – whether they come from people or states. Nick Thorpe has observed his trajectory as a leader not just over years, but decades.

The name of the artist is in the bottom right-hand corner of the magnificent wall painting which stretches along one side of the Delegation Hall of the Hungarian parliament. Painted in 1901, it depicts the newly crowned king of Hungary, Ferenc Jozsef, astride his horse on the coronation mound in June 1867.

The king was already emperor of Austria, and his coronation marked the launch of a new venture, known as the Dual Monarchy, whereby Hungary regained control of its sovereignty within the Austro-Hungarian Empire. In his right hand is a sword, and as part of the unique coronation ritual of Hungarian kings, he cuts the air in four directions – an ancient warning to Hungary's enemies to keep their distance.

An appropriate setting, then, for Viktor Orban's first meeting since his re-election with the media who will transmit or question his message for the next four years. At fifty, Viktor Orban is a solid, bull-faced fellow with a jovial manner combined with a frown which could freeze a lake.

I last met him in the gloomy corridors of the sports hospital a few months ago. 'Look – isn't that you-know-who?' my ten-year-old son whispered in my ear. And to my amazement, it was.

Flanked by a bodyguard and a doctor, he was waiting his turn on a wooden bench in a state hospital for a heart test, ahead of a minor operation. We went over to say hello. He introduced me jokingly to his colleagues as a representative of the 'hostile media', and wished my son well in his footballing career.

As Hungary's only charismatic leader since the fall of Communism, Mr Orban arouses anger and devotion in almost equal measure. 'In Europe today those who believe in a path beyond national interests and sovereignty are over-represented. The Hungarian government has a different viewpoint,' he told his Polish biographer Igor Janke. Europe can only be strong if it is made up of strong nations – that's Viktor Orban's creed. And the sword in his hand will cut anyone whom he suspects of challenging that vision.

I have watched him from close quarters since I first served him lapsang souchong tea in my Budapest flat twenty-six years ago. His strength has been his ability to adapt to new circumstances, and learn from his mistakes. His weakness, for me, is his almost messianic belief in himself as the saviour of the Hungarian nation. He has a fiercely majoritarian idea about democracy. The winner takes all, in his view, and opposition parties – what is left of them, after their second consecutive trouncing in the ballot booths – can join him in his crusade, or not.

At this point, my own path divides from that of many of his other critics, domestic and international. I am worried more by what he has *not* done than by what he *has* done in office. Hungary has been torn apart since the fall of Communism by a cold war between its liberals and national-conservatives. Ordinary people waited in vain for a compromise between the two elites – an agreement on the fate of Hungarian land, jobs, pensions and property. Viktor Orban, in 2010, was the first leader who had the power to impose a compromise – in the wording of the new constitution, for example – with which a maximum number of Hungarians could have agreed. Instead, he imposed a right-wing text which the opposition will probably spend the next hundred years trying to undo. The trenches in Hungary's political trench warfare get ever deeper.

In Europe, his Hungary will take a proudly national angle on each issue. In the economy, he believes in strong state intervention. On climate change, he is in the sceptics' camp, inclined to dismiss environmental protection as a conspiracy of the left.

'I dreamt last night of a circus,' the Hungarian writer Albert Wass wrote in 1947. 'And people, my dream was a monstrous one.' He goes on to describe nightmarish scenes, in which the Hungarians butcher one another in a circus, as the peoples of the world watch from their seats all around, placing their bets and cackling with laughter. All I can wish for my adopted homeland is that, with Viktor Orban or without him, there is a national theatre, not a national circus.

Hugh Schofield – France – Charlie Hebdo – 10/01/15

'Je Suis Charlie': a slogan and a sentiment which raced through social media across the world after the attack on satirical magazine Charlie Hebdo *on 7 January 2015. The title traded on its fearless, no-holds-barred satire; it was always ready to go for the jugular on any issue – from immigration to religion, feminism to free speech. It had courted criticism by republishing the Danish newspaper* Jyllands Posten's *notorious caricatures of the Prophet Muhammad in 2006, and in 2011 ran a deliberately provocative edition poking fun at Islamist politics in North Africa. After it published further cartoons depicting Muhammad in autumn 2012, the magazine was widely criticised and decided to take extra security measures. But no one had anticipated two gunmen bursting into its offices in Paris that day in January 2015 and killing twelve people.*

I have a memory of the first time I read *Charlie Hebdo*. It will have been sometime in the 1970s, and I was on a family camping holiday in the middle of France. How on earth it ended up in our possession I have no idea: my parents were certainly not the kind of people to read obscene political cartoons.

But I do remember what was in it. There'd been demos outside a nuclear plant that was being built and a protester had died in clashes with police. I have a clear recollection of the front page: the ultimate caricature of a brutish French riot cop, a grinning, bovine thug, in his hand the blood-dripping head of a long-haired hippy.

As a nerdy teenager at the time and distinctly damp behind the ears, I remember thinking: we don't get much of that back home! But it was vaguely stirring. What audacity! The image was so grotesquely exaggerated that you knew the message lay deeper.

They weren't just saying: we think the police are thugs. They were saying: we think the police are thugs, and to make our point, we are prepared to push to the limit any notion of taste, decency and accuracy. Because we can. Because it's funny. Why not?

Conservative types were shocked by *Charlie Hebdo*, and they were supposed to be. In those days its main target – apart from the police – was the Catholic Church. I've seen defecating popes, nuns in sex orgies, even nuns defecating on popes. *Charlie Hebdo* drew on an anti-clerical tradition in France that goes way back, and at some point the Church quite sensibly gave up complaining.

In my mind *Charlie Hebdo* merged with other childhood memories of France. Smelly loos at campsites; countryside that took your breath away; bosomy farmers' wives in patterned blue dresses; all that bucolic stuff and the chateaux – and then this blast of raw, confrontational anarchy. It was all part of the French mix.

Quite possibly the artist who drew that totally over-the-top picture of the French riot cop was the cartoonist Cabu.

That's another part of the mix – in France if you want to be taken seriously as an anarcho-agit-prop illustrator, get yourself a *nom de plume*. (In the obituaries this week, the cartoonists have all been designated by their nicknames – I don't think anyone cares what they are actually called.) Cabu was definitely around and drawing for *Charlie Hebdo* back in the 1970s. In his eighties he still dressed like he did half a century ago, and his mop of young hair over a humorous old man's face made him look like a cross between Ronnie Corbett and Elton John.

Anyway, Cabu's dead now. He was murdered. He was murdered for drawing pictures.

Could Cabu possibly have imagined in his wildest nightmares, when he set out on his career taking on the establishment sixty years ago, that his last seconds on earth would be that sudden noise at the door of the editorial meeting room; the incomprehension; the shouts; the shots? And then they say 'Which one's Cabu?' and the Islamist's Kalashnikov is pointing at your head.

That's how far the world has moved. Back when Cabu started, it was police and the pope. Now we have other things to worry about. And if there is one thing that everyone in the West frets about, it's Islam; it's Islamism; it's our countries' relationship with Islam; and it's our fear of what the future holds in a world where Islam – once our neighbour, once our enemy – is now part of us.

Cabu and the others knew this, and their reaction was to say: Well, if you're part of us, then think like us, be like us. Understand that there is a difference between mockery and persecution; that words and pictures are just that; and that part of the deal is that we rise above offence – yes, even when it's directed towards our own religion or beliefs.

Cabu would have been gratified by the outpouring of support on the streets of France these last dreadful few days. But he would probably also have said: Where were you all when we needed you? He and the others stuck their necks out for freedom. No one else did.

I miss the world of the anarchic 1970s, when the worst that could happen if you showed a copulating Christ figure was a letter in *Le Figaro* from 'outraged' of Aix-les-Bains. Now you die.

Joanna Robertson – France – Two Tribes – 02/06/16

The Revolution of 1789, the Commune of 1871, the student demonstrations of 1968: the streets of Paris have famously seen one wave of violent political protest after another. Well before the Gilets Jaunes ('yellow vests') movement emerged in October 2018, Joanna Robertson could see the effects of demonstrators' anger on her neighbourhood – rooted in some of modern France's deepest social and economic divides.

Josiane Bertrand has a small family business: a neighbourhood charcuterie selling sausage, poached pigs' trotters, pâté and jellied pig snouts. Her ham, she says, is the best in Paris and her queue of customers is long.

Despite the ceaseless rain outside – amongst all its other woes, France is now flooding – there's a convivial crowd waiting to be served – and the animated conversation is all about . . . strikes. If the opinion pages of *Le Monde* are to be believed, the charcuterie queue is a pretty accurate reflection of the mood of the country. I found it split, roughly half and half, between those for the Work Bill, and those against.

Philippe's twenty-eight. He's landed what most French would regard as a dream job. He's a *fonctionnaire* working in local government. A *fonctionnaire* is an employee of the French state in almost any form of public administration and service. That means a job for life – with solid pay and conditions, fixed working hours, a good pension, generous holidays. So what many young French people aspire to is not to change the world – explore, create, set up alone. Instead, with self-employment difficult and taxes punitive, they dream of becoming steadily employed bureaucrats with regular jobs.

Philippe knows he's lucky. And he's against any change. 'I'm happy,' he says. 'I know exactly where I am – and where I'll be in forty years' time, with a good pension.'

Eleonore, who has four children, is in her early forties. As a secondary school teacher, she's also got a job for life and generous state benefits. But, unlike Philippe, she's all for change. 'It can't go on like this. For every person like me, there are twenty or more with no hope at all.'

A quarter of all French people under twenty-five, many of them well-qualified, have no work. Many of them are from immigrant families, making their chances of employment even slimmer. These are the kind of people who voted François Hollande into the presidency in 2012 – with his pledge to end the country's employment troubles.

Now he's made a new promise, putting his whole political career on the line. He's not running for re-election next spring unless he cuts unemployment. A bold move for a president with an approval rating of only 14 per cent in a country riven by industrial disputes.

Along with his prime minister Manuel Valls (almost equally unpopular) and Pierre Gattaz – known as the 'boss of bosses', president of the largest federation of employers in France – Hollande stands against the combined power of the country's two biggest unions. The proposed Work Bill runs to over 500 pages. It aims to simplify and liberalise the French Work Code, which at 3,689 pages is a vast labyrinth full of perils for employers.

The unions won't even consider negotiations until the bill is removed from parliament. The president and his allies refuse to change a word of it. 'It's a good law, good for France,' says Hollande. The result? Total stalemate – an ongoing siege.

Just after one o'clock on Boulevard Montparnasse, the traffic disappears from the street. Cordons of riot police move in, three columns deep, flanked by armoured vans. There's a whirr of helicopters overhead. In the distance, a gathering roar and blare – the protesters. The noise becomes deafening. The riot police take up positions.

On the glassed-in terrace of a popular restaurant overlooking the boulevard, Frederic the waiter temporarily locks the doors – and those having lunch find themselves exhibits in a kind of transparent gastronomic showcase, along with various grilled fish, bottles of wine, and assorted desserts.

Looking in from the outside, hundreds of protesters passing down the boulevard – some marching, others ambling, a few dancing to the music booming from the accompanying floats.

Looking out from the inside, the lunchers. They comment on the demonstrators; the demonstrators wave cheerily at the lunchers. There's generally resigned, gently amused talk amidst the eating – 'Here we go again' and 'Where will this round end?' and self-deprecating comments along the lines, 'We French do love to demonstrate . . .'

Then it all subsides, passes on – the noise, the marchers, the red balloons and pounding music – taking the helicopters with it, leaving a trailing wake of litter. Frederic unlocks the doors. The conversation leaves the political, returns to the personal.

Russia

Steve Rosenberg – The Jigsaw Puzzle – 21/05/11

Who can know the mind of Russia's leaders? The very inexact science of 'Kremlinology' – reading the true intentions of powerful men in Moscow from the limited information which is allowed to reach the public in Russia, let alone the West – stumped many an analyst throughout the twentieth century and well into the twenty-first. As Vladimir Putin and his ally Dmitry Medvedev appeared to be swapping power between them in 2011, Steve Rosenberg tried to figure out what was real competition, and what was just choreography.

I've often thought that trying to make sense of Russian politics is a little bit like trying to do a 5,000-piece jigsaw with half the pieces missing and with no picture on the box to help you. You can spend an eternity struggling to fit together the bits you have – and just when you think you've uncovered a connection, a tiny clue to the mystery, you realise, no, that doesn't fit after all, and you have to pull it apart and start again.

Because politics in Russia is conducted behind closed doors (with the curtains drawn and the shutters down) most of the jigsaw pieces that Kremlinologists have to work with are hints, rumours, speculation – plus whatever public announcements the politicians deign to make.

Since returning to Russia last autumn, I've been poring over this puzzle, trying to piece together what's happening here and what's going to happen at the next presidential election in 2012. And I have to admit, it's had me stumped.

The problem is that the picture keeps changing. One minute, it looks like Vladimir Putin will replace Dmitry Medvedev in the

Kremlin. The next, that President Medvedev will get a second term. Sometimes it seems that Putin and Medvedev are one team, sometimes that they're rivals. And sometimes – that the next president of Russia will be someone completely different. Make sense of all of that!

But this month two things have happened which may hint at how the puzzle will be solved. Dmitry Medvedev held his first high-profile news conference since becoming president more than three years ago. The event was broadcast live by three Russian TV channels; it was the perfect stage for a major announcement about his future. And that's exactly what the hundreds of journalists packed into the hall had been expecting. In the end there was no big announcement; in two hours the president said very little that was new. Meanwhile, Prime Minister Putin has created a new political movement, the All Russian People's Front. It aims to unite different parties, unions and business groups around the prime minister. The way the jigsaw pieces lie now, it looks like Mr Putin will make a play for the presidency.

If that really is the final picture of the political jigsaw, what do Russians think about it? Mr Putin has fierce critics, particularly among the more liberal pro-Western section of society. They accuse him of rolling back on democracy, eroding human rights and encouraging corruption. These are the kind of criticisms you often hear in Moscow.

I travelled to Krasnodar in southern Russia – far enough from Moscow to feel the pulse of the nation. Most of the people I spoke to in the city and in the villages outside expressed support for Mr Putin. Like Irina, a twenty-four-year-old marketing manager, who I chatted with in the park. 'Speaking as a woman, Putin's more handsome than Medvedev,' Irina told me. 'What's more,' she continued 'it's Putin who began the reforms. Medvedev's just continuing them.'

A man called Pavel told me that Putin was a 'tough guy – just the kind of leader Russia needs'. Pensioner Olga said she thought that Putin was 'closer to the people' than Medvedev. Curiously, everyone had something to complain about: the state of the Russian education system, low pensions, rising petrol prices. But few people blamed central government directly for their problems, preferring to hold local officials or multibillionaire oligarchs responsible for them. 'It's not Putin's fault that fuel prices are going up in our city,' a man called Sergei assured me. 'How could he know what's happening here? The local bureaucrats hide the truth from him.'

I wasn't the only visitor to Krasnodar that day. Vladimir Putin was there too, on a trip to promote Russian sport. At the local physical education college, he watched young people wrestling and boxing and lifting weights; Russian TV cameras crowded round to capture the action. Putin loves being associated with the world of sport. He's often shown on TV here doing judo, skiing, swimming, or playing ice hockey. It boosts his image as a strong leader. I suspect it's one of the reasons he remains Russia's most popular politician.

In the corridor of the college, I spotted Mr Putin's spokesman Dmitry Peskov. This was a chance to fill in key pieces of the jigsaw and find out whether the picture I think I've formed is accurate. Was Vladimir Putin planning a Kremlin comeback, I asked?

'I don't know who's going to be the next president of this country,' Mr Peskov said with a large smile, followed by a little chuckle.

Oh well, it was worth a try. I'll just have to keep puzzling it out.

Lucy Ash – Putin and the Patriarch – 29/09/12

During the Soviet period, the USSR was atheist as a matter of policy. Although Russia is still officially a secular state, its Orthodox Church increasingly portrays itself as intrinsic to national identity – and its politicians are inclined to agree. Some argue that the Church's political influence is greater today than at any time since the seventeenth century. Patriarch Kirill has made no secret of his strong support for Putin, praising his leadership as 'a miracle of God'. He also described the notorious, if very short, performance of a 'punk prayer' by the feminist band Pussy Riot in Moscow's Christ the Saviour Cathedral as part of an assault on the nation by 'enemy forces'. As Lucy Ash visited the president's favourite monastery, on Europe's largest freshwater lake, a bill criminalising blasphemy was going through the Duma. It declared 'public actions, clearly defying the society and committed with the express purpose of insulting religious beliefs' to be a federal crime punishable by up to three years in jail.

'Now there are fewer boats coming in – thank God we'll get peace and quiet.' End-of-season talk from a worn-out hotel manager, you might think. But Mikhail Akulinovich isn't a typical hotelier. He wears a black habit instead of a shirt and tie, and these days he's known as Father Matvei. His hotel isn't typical either, housed as it is inside the fourteenth-century Transfiguration Monastery on the rocky island of Valaam in Lake Ladoga.

Father Matvei rises every morning at half past four and spends three hours in church before checking his bookings, and welcoming new guests. Despite its remoteness, 120,000 pilgrims, tourists and VIPs flock here each year.

I wondered why he had come here after spending most of his life as an airline pilot. He fished under his habit and pulled out a long, slightly greasy ribbon embroidered with quotations from Psalm 91, 'my refuge and my fortress'.

'My mother worried about my safety and wanted me to wear a cross,' he said, 'but I didn't want to. We pilots had to strip off for medicals and in those days we were all supposed to be atheists.' So Father Matvei's mother stitched the psalm through the ribbon and sewed it into the lining of her son's Aeroflot jacket. He wore it just to humour her – until one day the engines on his small plane cut out. 'I remembered the ribbon and silently prayed. Somehow the co-pilot and I managed to land and we walked away without a scratch. If I hadn't prayed, we'd have died and might have killed people on the ground too.'

After the collapse of the Soviet Union in the early 1990s, he began attending church, teaching the Bible to children and going on pilgrimages. Then five years ago he left the airline and his wife to join the brothers on Valaam. His wife, herself a churchgoer, raised no objections, but not every monk on the island has such an understanding family.

His beard and round glasses may make Father Iosif look like a St Petersburg intellectual but when he speaks English he sounds like a native New Yorker. At sixteen, he decided he wanted to become a monk. His father, who ran a furniture company, was horrified and packed him off to business school in the USA, hoping to knock some sense into him. 'As you can see,' said Father Iosif, 'that didn't work.' His mother and father have now embraced Orthodoxy and recently got remarried by a priest. 'In the old days, parents used to drag their children to church; now it's the other way round,' he said.

Young monks like Father Iosif embody the renaissance of Russian Orthodoxy after almost a century of Communism. This place houses a small copy of the Mother of God of Valaam, the monastery's most famous icon. An inscription underneath notes that it was sent into space and orbited the earth 488 times. Whose idea was that? I asked. Father Iosif suggested the decision was taken at the very highest level and mentioned that President Vladimir Putin is a frequent visitor. 'He is our benefactor ... a person who provides aid in all possible ways to our monastery.' Is the president a holy man, I wondered? 'Only God knows that,' said Father Iosif, beginning to sound a bit tetchy. 'Why are you so interested?'

Patriarch Kirill, the head of the Russian Orthodox Church, has his own residence on the island on a pine-clad promontory next to the newly built church of St Vladimir. Many Russians worry that the current patriarch is leading his flock closer to the authorities rather than God. Some have even likened the Orthodox Church to the Communist Party – a mechanism to stifle dissent and force obedience to an authoritarian regime.

The jailed members of Pussy Riot say their performance on the altar of Christ the Saviour Cathedral in Moscow was prompted by illegitimate elections and Patriarch Kirill's endorsement of the current president. One of those young women, Maria Alekhina, is herself an Orthodox believer. From her prison cell she wrote: 'I thought the church loves all its children, but it seems it only loves those who vote for Putin.'

Paul Henley – A Handshake for Homophobes – 02/03/13

One feature of the public discourse in Russia in the last decade has been its increasing stress on 'family values'. There's been much talk of the 'natural authority' of fathers and husbands, reduction of penalties for domestic violence, increased awards and grants for the parents of large families, and a new emphasis on 'traditional' gender roles. According to this worldview, growing acceptance of LGBT people elsewhere is a mark of modern decadence. In the summer of 2013 the Duma unanimously passed a measure to make the 'distribution of homosexual propaganda to minors' an offence punishable with large fines (or deportation, for foreigners). Far from Moscow, Paul Henley got plenty of insight into the prejudices at large.

Giving your full concentration to unrelenting vitriol can be quite tiring. My fault, I suppose, for actively seeking it out. It's just that after many hours of holding a polite expression and nodding while being told that you are (take your pick) disgusting, contaminating or a threat to civilised society, then what you really want is several drinks, a fight, or a darkened room and a comfy bed.

I'll settle for a fold-down bunk on a twelve-hour night-train to Moscow, but I can't pretend there aren't issues. Persecution complexes are all about context. The black-and-white striped mattress doesn't help. Or the screeching of brakes that make it sound as if we're being shunted into sidings. Or the shouts in an unfamiliar eastern European language and torchlights on snow-covered station platforms. Don't underestimate the effects of repeatedly hearing how people like you should be 'rounded up and punished'. The harmless ticket inspector knocking at the door can seem surprisingly threatening.

In the cold light of a Russian day, I can see that not everyone is really out to get me. The dangerous ones think that gay people exist only in silver knickers anyway, waving banners and dancing to Madonna.

The Cossacks of Lipetsk were something, though . . . historically, specialists in pogroms, turning up in their capes and bearskin hats and knee-length boots and saying homosexuals weren't even worth the effort of using their steel-tipped whips. One of them made a spitting sound when he said that, unlike the English, they knew how to deal with perverts. They posed for photographs with me, all smiles and arms round my shoulders – a moment of undercover revenge.

Of the many people I've met in the past week who happily told me they hated homosexuals, I don't think any would admit to having met one. 'I've seen them on TV – so I know they exist,' huffed one lawyer and dedicated drafter of legislation to curb minority rights.

Such wilful ignorance and hyperbole I'd expected. What I found more interesting in my travels around homophobia was the idea, among those who thought of themselves as open-minded, that gay people were tolerable ... as long as they kept their gayness entirely secret, never referred to it in conversation, never admitted to having a partner – God forbid children – and kept up a lifelong pretence of being heterosexual.

I was constantly told that sexual expression was a private matter in Russia and that conservative family values ruled. But I wondered how this tallied with the boy–girl couples I was getting used to seeing putting their tongues in each other's mouths at bus stops, or with women's clothes moulded to enhance the curve of bottom and breasts even at -15°C. In one small southern town, the hotel where I'd been recommended to stay had tiger-print polyester sheets, soft porn on the walls and graphic noises from the next-door rooms. Its restaurant, on the town's main street, was due to be the venue later that week for 'Sweet Couple' – a live erotic show billed as 'shocking' – to celebrate Russian Army Day.

So, under the new law against 'homosexual propaganda', gay people in Russia could theoretically be arrested for holding a social in a café. But it was apparently OK for straight people to have full-on sex on a café table.

Russia's full of reasonable people, too. You don't get the full picture from the extremists alone. But this law's still popular enough for a crowd of 1,500 to turn out to intimidate fourteen human rights demonstrators who voiced their opposition to it in the central square in Voronezh last month. One of them got beaten up – a softly spoken, slight boy called Pavel who says it took him years of gathering courage, through a suicidal adolescence, to come out.

It had been a long week of being journalistically impartial.

I asked the leader of the mob, a man fond of military uniforms and crucifixes from the All-Russia People's Union, if *he'd* ever met any gay people. He laughed and said no – he didn't keep any 'damaging connections'. He said that in his organisation, they joked that if you shook hands with a pervert, you became one. I gave him a nice firm grip when we said goodbye. I'd taken my glove off specially.

Bridget Kendall – Solid and Confident – 05/06/14

As a fluent Russian speaker who was the BBC's Moscow correspondent during the hectic news years of 1989 to 1995, Bridget Kendall can take the deep as well as the long view of Russia's current global status. She reported on President Gorbachev's attempts to reform the Soviet system with perestroika; the break-up of the vast USSR into fifteen independent countries; the attempted communist coup of 1991 and Boris Yeltsin's rise to power. Revisiting the Russian capital in summer 2014 she found a nation apparently bursting with confidence in itself and in its leader.

Moscow was in the midst of a heatwave when I arrived. The heady scent of lilac on the long warm evenings brought people out onto the streets. One night on the embankment of the Moskva River, a drunken band of students tried to engage me in conversation, and then staggered off down the street.

There is always a slightly celebratory air at this time of year – a realisation that winter is over and a few precious months of sun are here. But this year's euphoria is underpinned by a sense of having proved to the world that Russians can come out on top. A surge of self-confidence, after the humiliations of post-Soviet collapse.

'We're on a roll,' one Russian TV editor in chief told me. 'No one can stop us now.'

'First the Sochi Olympics,' said another man, who I met on Red Square, 'then we got Crimea back. And now we've won the World Ice Hockey Championships as well!' It's as though everyone is a spectator at a sporting event, or watching a game show on TV – a detached enjoyment, as though this orgy of Russian patriotism carries no consequences, so there is no point thinking where it might lead. To be fair, lots of people *are* watching events unfold on TV.

The sheer volume of different state-controlled channels is overwhelming. All glossy and well-funded, each with their own slightly different but ultimately similar version of the news: hours of slick, punchy and emotionally charged reports, set to thumping music, making your heart race. On Ukraine there's one message: that the violence in eastern Ukraine is all Kyiv's fault, that Ukraine is crawling with Russian-hating neo-Nazis and fascists, and the US government is fuelling the crisis behind the scenes while Russia tries to act as peacemaker.

'Aggressive and deceptive propaganda ... worse than anything I witnessed in the Soviet Union,' was the verdict of Lev Gudkov, the director of the Levada Centre, Russia's most well-respected polling organisation.

Cultural historian and publisher Irina Prokhorova went further – she called it Stalinist, reminiscent of the anti-Western hysteria which marked the grim repressive years of the late 1940s.

What worries the liberal intelligentsia is that the aggressive tone is being directed inwards as well. Ever since President Putin warned he would not tolerate a fifth column of national traitors in his historic speech on Crimea's annexation in March, Russia's beleaguered opposition has felt under siege.

'We've been investigated four times this year already,' said Lev Gudkov. 'We live on tenterhooks. At any point they could close us down.'

'Be careful what you write about us,' whispered another academic. 'You understand how it is here now.'

So what or who is fuelling this aggressively anti-Western mood?

Up on the Sparrow Hills, on the fourth floor of the sociology department of Moscow University, is a room marked 'Centre for Conservative Research'. This is the office of Professor Alexander Dugin, who welcomes sanctions because he wants Russia to split with the West. He also thinks President Putin should and will invade eastern Ukraine.

Once a fringe figure, he is now seen as the ideologue at the heart of Russia's new conservatism. His long grey beard makes him look like Solzhenitsyn – another Russian thinker who wanted to reunite all Slavic-speaking lands. But Alexander Dugin speaks the language of postmodernism, not religion. And his focus is politics not spirituality. On his desk among the books and half-drunk coffee cups are not icons, but a roll of the black-and-orange-striped ribbons used by rebels in eastern Ukraine to denote their loyalty to Russia. Is he in touch with these separatists? They are his friends. And does he advise the Kremlin? Has he met President Putin? He smiles slyly. 'That's personal,' he says. 'I'm not going to say.'

The question is how long this new patriotic conservative mood in Russia will last. Publisher Irina Prokhorova wonders if it could be the last gasp of an authoritarian regime under threat, like Stalin in the late 1940s. But the pollster Lev Gudkov thinks the surge of enthusiasm for President Putin will endure long enough for him to get re-elected in 2018, which means – if he stays in good health – Vladimir Putin could well be Russia's leader until 2024. At least.

Steve Rosenberg – Russia – Siberian Chill – 27/02/16

While the basic principles of journalism haven't altered – the audience still wants to know the who, where, when, what and why – the process of turning reporting into reports has certainly changed over the past decade. One striking trend has been the ratcheting-up of scrutiny of foreign correspondents themselves. BBC journalists are still expected to be objective and dispassionate – but now they're also answerable to more or less everyone, all the time. Not just to BBC news managers, or the long arms of state security where they report; their steps are now also followed by pressure groups, online observers and, of course ... other journalists. Which can get awkward, as Steve Rosenberg relates ...

When morning breaks over the River Ob, the commuter trains rumbling across the railway bridge disappear into clouds of mist. They are the ghost trains of a Siberian dawn. Through a pink-and-orange sky bursts a golden sun – and the Ob is bathed in light. But it is bitterly cold. What could be colder than a Siberian winter? I'll tell you what: the icy reception we're about to receive in Novosibirsk.

We're out on the streets filming a report about the Russian economy. Suddenly I hear a voice: 'I'm from Novosibirsk Television. Are you the BBC?'

The reporter from this pro-Kremlin channel – he's called Alexander – has tracked us down and his cameraman is already filming me.

'The journalist community of Novosibirsk wants to know why you are here! What kind of film are you making?' he asks. He informs me that a recent BBC programme on alleged Kremlin corruption was a 'pack of lies ... a provocation! Are you making the same kind of film here?' he barks.

I tell the journalist that we're making a news report about life outside Moscow – is that a problem?

'Yes, it is,' he replies, 'judging by the kind of programmes the BBC makes.'

In the hotel that evening, I switch on the TV to see what kind of programmes Siberian Television makes: rather imaginative ones, judging by its main evening bulletin. 'BBC in Novosibirsk' is headline news. But it's more fairytale than fact. The newsreader introduces us – incorrectly – as the same BBC team that made that programme about the Kremlin.

'Now they've received a new assignment straight from BBC headquarters!' says the reporter, as if this is some kind of spy thriller. The reporter says we've been filming things that we haven't. And, when the report ends, the newsreader announces: 'Tune in tomorrow to find out what the BBC's report is really about.'

Next morning we drive to a business centre to interview the head of a local internet company about Russia's recession. And who should turn up, uninvited? Yes, it's Alexander again, with cameraman and questions.

'I want to know, what kind of a report about Russia are you making?'

By now it's clear we're being followed. I tell Alexander we're not used to being pursued.

'I haven't been following you,' he replies. 'I just got a call telling me you're here.'

I tell Alexander I don't think it's very polite to barge into someone's private office.

'But we're doing the same job,' he replies. 'You're making a report about Russia – and so am I.'

That night I sit down to watch the local news and there we are again. Not surprisingly, the part of the conversation that mentioned being followed and barged in on didn't make it into this report. Instead Alexander declares, sarcastically, that 'The BBC knows how to show the "pretty side" of Russia.' He accuses us of filming 'food markets, bus stops and private houses'.

'We may have plenty of dirt in our city,' he says, 'but there's plenty of dirt in the West, too.'

In a bizarre attempt to prove his point, he shows some photographs of dilapidated homes in the Essex seaside town of Jaywick: officially the most deprived neighbourhood in England. 'The BBC would never show this,' he declares.

I'm not too surprised by Siberian TV's frosty coverage of our visit. As relations between Russia and the West have deteriorated, the amount of anti-Western reporting on pro-Kremlin TV channels has risen. The Russian authorities, too, often complain about what they see as increasing anti-Russian rhetoric in the Western media.

But I wonder how effective this message is in Russia. Next day on Lenin Square I interview some pensioners who are protesting against the government's economic policies. Suddenly one woman says to me: 'Hang on, I saw you on TV last night. You're that British journalist, aren't you? You're bad!'

'Do you believe everything you see on pro-Kremlin channels?' I ask. 'You've come onto the streets to criticise your authorities. So why are you so ready to accept that the BBC is bad?'

'Well, it's all right for *us* to criticise our government,' she replies, 'but if we complain to *you* about our government . . . well, that's not patriotic.'

Ukraine

David Stern – My Friends on the Barricades – 22/02/14

'It's like Game of Thrones *on ice – but for real.' The surreal scenes of something like mass siege warfare on the snowy streets of Kyiv in the winter of 2013–14 left many an observer half-elated, half-appalled. During Ukraine's 'Orange Revolution' a decade earlier, mass civil disobedience had forced the rerun of a presidential election, bringing Viktor Yushchenko to power despite an attempt to assassinate him via dioxin poisoning. But his old rival and competitor in the 2004 contest, Viktor Yanukovych, would win in the next round in 2010 to become the fourth president of independent Ukraine. By 2013, Yanukovych's government was drifting ever closer to Russia, and growing ever more resistant to the EU and its demands. Moreover, public resistance was growing – and young people in particular gathered in Kyiv's Independence Square to voice their discontent. These 'Euromaidan' protests grew more turbulent and more dangerous; unlike the bloodless Orange Revolution, they saw over a hundred deaths. David Stern began to worry about his sources' safety.*

Nearly every day for the past three months, I have passed through the protest camp, whether I was working on a news story or not. My clothes grew saturated with the smoke from the activists' campfires. If I was feeling a bit hungry or cold, I could grab a quick snack or cup of tea from one of the many volunteer stands handing out free food.

When I'd go to the post office to pay my bills, the main lobby would be packed with protesters, tired and sooty, catching a nap on the chairs or slumped against the wall. As I did my shopping,

31

there were always activists in front of me, stocking up on food, or sometimes vodka.

As the stand-off continued and violent clashes broke out, the streets around my flat were transformed by a carapace of makeshift barricades, which seemed to spring up like mushrooms. One day the street was clear. The next, activists were standing next to piles of snow and debris, directing traffic and throwing gimlet stares at men they thought looked suspicious.

Of course, I never expected to be in the middle of Ukraine's biggest post-independence crisis when I moved here five years ago. Living next to a combat zone has been strange, to say the least – especially one contained in a compact area in the city's centre.

Now the turmoil has turned deadly. My wife and I sit glued to the television, with its images of the fighting just blocks away from us, the bodies laid out on the square and the angry clashes in the regions. As the number of dead climbs, and climbs, we look at each other silently, not believing what is taking place in our lovely city. Often, my thoughts turn to the many activists I have met.

One in particular, a member of the ultra-nationalist Right Sektor, has been on my mind. We struck up our acquaintance in December, as I was watching riot troops mass close to my apartment, just before their first big push to try to clear the square. He told me his name was 'Frank' – a pseudonym. Unlike many of the right-wingers, he was small and wiry. When he took off his ski cap, his head was shaved – a disguise of sorts, he said, to hide from the security services.

We spent the next few hours walking around Kyiv, looking for other locations where the riot troops might be massing. As we walked, Frank told me about the right-wing movement. This was before the political crisis had turned lethal. He said the ultra-nationalists had stockpiled weapons and would unleash a partisan war if police moved them on. At the time, I thought this was all braggadocio. Now I'm not so sure.

Many questions have been raised about whether the right wing has initiated clashes with the police. I can't answer this. What I can say, though, is once the fighting erupted, far-right activists have been passionate participants. They have also played an oversized role in the protest camp, which itself has been an influential, but not decisive, force in the general anti-government movement. As the crisis has worsened, downtown Kyiv has been inundated by scores of ultra-nationalist fighters. They march in formation, with wooden clubs and metal rods, wearing helmets and ski masks, and shouting, 'Glory to Ukraine!'

But the right is a small part of the Euromaidan protest movement, which itself unites a wide range of groups – including Jewish and LGBT activists. All ideology has been forgotten – or muted – in the single goal of removing President Yanukovych.

Frank himself seemed a bit unclear of what his group, the Patriots of Ukraine, stood for. 'They say we are fascists and anti-Semites. You don't think that, do you?' he asked. In fact I did think this – or something close. I've seen the group's website and spoken to other activists.

There are some who say the right is experiencing an upsurge only because of the loathing and rage of many for President Yanukovych. Once he is removed, they say, the ultra-nationalist rhetoric will deflate. That may be so – but it is a question for another day, once peace is restored to Ukraine. Right now, I think about Frank. And hope he is OK.

Steve Rosenberg – Post-Revolutionary Daze – 01/03/14

After months of turmoil in Kyiv, the end, when it came, was quick. President Viktor Yanukovych left Ukraine in a hurry on 21 February 2014, and very next day the Ukrainian parliament summarily voted to strip him of his post, as he 'was not present to fulfil his responsibilities'. A warrant was issued for his arrest on the grounds he was legally responsible for some of the deaths during the Euromaidan protests. Since then he's been found guilty in absentia, by a Ukrainian court, and sentenced to thirteen years' imprisonment for treason. Steve Rosenberg had to hurry to keep up.

I've been reminded this week that when change happens, it can happen very, very fast. When I woke up last Saturday, Viktor Yanukovych was still Ukraine's president, and his arch-rival Yulia Tymoshenko, the former prime minister, was still a prisoner under police guard in a hospital east of Kyiv.

By the evening, everything had changed: Ms Tymoshenko was up on stage in Independence Square in Kyiv making fiery speeches from her wheelchair; Mr Yanukovych had been removed from office by parliament and was on the run; and I was strolling through his back garden, marvelling at the opulence of his mansion, private golf course and ostrich farm, and checking out the jacuzzi in his luxury spa. What a difference a day makes in Ukraine.

I will never forget the moment the gates to Viktor Yanukovych's presidential residence swung open and the public poured in to look round. This was not an angry crowd baying for blood and revenge. People were just curious to see how their president had lived. Strolling through the vast grounds were mums pushing prams, dads with children on their shoulders. People posed for photos by the mock Roman ruins, next to cages with pheasants and in front of a giant galleon – Mr Yanukovych's multimillion-pound banqueting hall boat in a garden lake. 'So THIS is what they've spent all our tax money on!' one woman said to me, shaking her head in disbelief.

On guard in Mr Yanukovych's sauna were members of the Independence Square Self-Defence Force. 'No marks on the furniture, please!' they barked, to stop visitors trying out the plush three-piece suite near the plunge pool. 'We don't want anything damaged, you know, we don't want the world to think that we're hooligans.'

It seemed such a remarkable thing to say when you consider the horrors this city experienced just two days earlier. That was when dozens of people – most of them protesters – were shot dead by pro-government forces right outside our hotel. Soon after the violence broke out, the hotel foyer was turned into a makeshift field hospital and a morgue. Corpses were laid out near reception. They were covered up with hotel bedsheets and priests chanted prayers for the dead.

That afternoon, after the shooting had died down, a medical volunteer called Alexander came up to me. 'I want to show you something outside,' he said, 'so that you understand what war really means.' We left the hotel, walked round the corner onto Institutska Street and stopped. 'You see that black hat lying on the ground?' Alexander said. 'Well, there's a sliver of someone's brain inside. I'm going to collect it and try to match it to one of the bodies back at the hotel. We have to return it to the right person. Because that was a piece of someone's mind, someone's intellect: that was someone's thoughts.'

I think Alexander was in shock. Probably, so was I. At that moment I just couldn't believe I was in Kyiv. That this nightmare was unfolding in what is one of the greatest cities in the whole of the former USSR. A city of chestnut trees and sweet-smelling coffee shops, of river cruises and concert halls; a city I'd fallen in love with back in Soviet times when I'd lived and studied here for three months.

It was this violence that triggered change. Whether or not Viktor Yanukovych personally gave the order for live rounds to be used on the protesters, the bloodshed sent his administration into a tailspin. He claims his removal from power is a coup. But even his own party had blamed him, disowned him and denounced him as a coward for running away.

But what now for Ukraine? The country has massive debts and requires urgent financial aid. It faces the threat of separatism, too. This week armed men seized administrative buildings in the Crimean Peninsula – a part of Ukraine, but one with an ethnic Russian majority. Crimea intends to hold a referendum on greater autonomy from Kyiv. There is suspicion that, behind the scenes, Moscow is pulling the strings.

Ukraine does now have a new government, a new prime minister – a new caretaker president, too, until elections in May. But there will be no honeymoon period for this administration. The protesters I have spoken to on Independence Square tell me they are deeply suspicious of everyone in positions of power. They believe that it's the people, not the politicians, who have forced change in Ukraine.

For now they seem determined to stay on the streets to make sure their new government doesn't deceive them.

Fergal Keane – War Made Itself at Home – 23/05/15

The escalation of tension between Russia and Ukraine continued – and erupted into a shooting war by the spring of 2014. Russia took control of the Crimean Peninsula and fomented anti-Kyiv feeling in the east of Ukraine, where many Russian speakers live; Ukraine mobilised thousands of volunteers to try to hold what it could of the rest of the country. The conflict metastasised, and a year later Fergal Keane, who'd reported on many other wars elsewhere, recognised how it had become the 'new normal'.

The soldier was red-haired, with a thick beard, a girth that spoke of many good dinners, the flushed complexion of a man who lived well. Although our time with him was short, a brief trip to the front, he was one of those people in whose company you instinctively felt safe. He smiled when he told us that we'd be fine. And he meant it.

A few days later he was dead. A landmine did for him on the front. The war consumed the life of the red-haired soldier and took him from his wife and four daughters.

It is one year since I first travelled to the east. I came through the gleaming airport at Donetsk into a steadily deteriorating political crisis. There were guns and men with balaclavas at the city's main administrative building – but there was a veneer of normalcy. Couples strolled by the Kalmius River in the evening, passing old men doing stretching exercises, and the old women watching their grandchildren play in a sandpit.

But how quickly war made itself at home here. Since last summer trenches have been dug, armies deployed. Nightly the rockets and shells erupt in the contested districts. Tens of thousands have fled.

I remember the shock of the early civilian casualties here. In an apartment block near the railway station a Ukrainian army shell landed in the playground. Two people, passing in the wrong place at the wrong time, were cut down. The blood was drying when we arrived; a shoe belonging to one of the dead lay near the impact point of the explosion.

There is little shock any more. How could there be, really, after the shooting down of flight MH17, and all those people lying in the burned field, twisted and broken in the wreckage, or hurled across the valley when the missile struck?

That first morning on our approach we came across a small figure - a local had covered the infant with a sheet – that had been blown far away from where the rest of the bodies lay, among them those of the child's parents. I remember berating a rebel officer for leaving the bodies of the dead uncovered. I will never rid myself of the image of one young man, lying naked by the roadside while the media, the international observers, the rebels, the locals, all just passed him by. These are just fragments of the last year of war.

A few weeks ago, I walked around the remains of the airport terminal – fearfully, tentatively, for both armies shell and snipe regularly. My colleagues spotted the unburied, headless corpse of a soldier lying by the side of the road. He had lain there for weeks. Many more were buried under the rubble.

Vanya Marchenko was thirty years old when he was drafted and sent to fight at Donetsk airport. He was killed in January. I met his mother Leonida a few days ago at her home on what was once a collective farm outside Kyiv. She is bitter. What did her boy die for, she asked me? It was the children of the poor who died, while the rich kept their privilege. A year into the war and this is a familiar refrain among the overwhelming majority who live outside the small circle of the moneyed and well-connected.

Ukraine these days is a mess. It teeters on the edge of economic collapse. Corruption is rampant. The government doesn't have the military power to defeat the rebels, who are armed, trained and bolstered on the ground by Russia.

The rebel areas are ruled by the gun: a conglomeration of militias, armed personality cults, criminal gangs and the influence of the Russian secret service.

The world is preoccupied with other crises, and far too divided politically to solve the problem of Ukraine. After all, the violence is at a low level. Soldiers die, but most civilians have fled from the frontline areas. There is no nightly drama and horror from the east on our television screens. The ceasefire is a fig leaf that allows the world to say there is a working peace process.

But the fighters don't believe it. Ask them about peace, on either side of the frontline, and you hear the same thing: not yet. And then ... frequently ... laughter.

Financial Crisis

Sarah Rainsford – Spain – The Indignant – 06/08/11

Even in hindsight, the size, the risks and the consequences of the European financial crisis defy any easy analysis. The dizzying figures which punctuated every news bulletin – the billions of euros of value wiped from the value of stocks and shares; the graphs showing plummeting savings, or rocketing unemployment; the charts with ever-shrinking pies showing how much less national governments had in their newly austere budgets – were often hard to fathom. And if even the so-called experts – the technocrats and bankers and politicians – hadn't foreseen the risks of economic disaster, how was the average person in the street supposed to manage? Sarah Rainsford found that in Spain, the public's patience did have its limits.

If you walk down Gran Vía in Madrid, you can still see the traces of a recent protest by the indignant. There are bright red handprints on government buildings, paint smeared over cash machines and on one bank's wall, the word GUILTY in big, black letters. It's quite clear, then, whom protesters here blame for the economic crisis and the fact that 46 per cent of young Spaniards are unemployed. They're angry at the banks, whose risk-taking started this recession, and at politicians, who've responded with spending cuts, privatisations and redundancies.

Back in May, the 'indignation' of Spain's young generation finally spilled out onto the streets. Thousands pitched a protest camp right in the heart of Madrid. Every night, Sol Square was filled with huge rallies. The camp has gone now. But this 'Spanish Spring' some had predicted, the spirit of May, has survived into the long, hot summer.

Recently I went hunting for a group of the protesters as they marched on Madrid. We finally met, after a flurry of text messages, just outside the pretty town of Aranjuez. Slumped by the roadside, the young marchers were exhausted. 'You don't sleep or eat properly, your muscles ache. Some days, you just collapse and think you can't go on,' Carolina admitted, after twenty-three days' walking from Granada. Hers was the southern march, just one of seven columns converging on Madrid.

Carolina joined in because she's highly educated and unemployed – and angry at what's happening to her country. She told me the pain and effort were worth it. 'We want the politicians to see that millions of people are unhappy,' another girl explained as her friends sucked on watermelon chunks and gulped down water.

The marchers wanted to be noticed. They stopped in towns and villages on their way to hold assembly meetings in the open air, spreading their own message and collecting local concerns to carry to Madrid. As they entered Aranjuez, they filled the high street, dancing and singing. 'We won't pay for your crisis!' they'd chant, or: 'They call this democracy, but it's not!' A few broke away and surged, still chanting, into banks – very passionate, but peaceful. All along the road, shoppers and shopkeepers stopped to watch. The protesters were a bedraggled bunch but almost everyone I spoke to was very supportive.

Like Santiago Montalba, who owns a jewellery store in Aranjuez: a family business, on the verge of collapse. He told me sales have slumped by 80 per cent. Three of his four children are unemployed. 'These people have studied, prepared for life and they're in a dire situation. Things can't go on like this,' the jeweller told me, as the indignant poured noisily past his shop. 'We have to support them. They're the only ones with the power to try to change things.'

Another woman in her sixties told me she's a great fan of the protesters. She's defaulted on her mortgage and the indignant have promised to help stop her house being repossessed. It's a cause they've taken up across the country, deploying flash mobs to physically prevent police and bailiffs from evicting families on behalf of the banks.

That had an impact. The government has brought in new

regulations, protecting a larger slice of people's income from banks reclaiming mortgage debts. Santander Bank has declared a three-year moratorium on repayments for mortgage holders whose income falls significantly. A spokesman said the bank was 'sensitive' to people's difficulties.

But the indignants' demands go broader and deeper. They want Spain's electoral system overhauled and tougher sanctions against corruption, and they want politicians who won't bow to every whim of the financial markets. Many Spaniards call this a social awakening. They're energised, enthusiastic – excited, even. 'We didn't know each other before this,' Aurora told me at a weekly assembly meeting in a Madrid park. 'Now we're talking, we see how much we have in common. And we can fight for change.'

The paint splashes on Gran Vía suggest there are more radical elements within this very diverse movement. But the vast majority are peaceful, intent on reforming the system, not destroying it.

Recently they got some high-profile encouragement when a Nobel prize-winning economist dropped in on an assembly. Joseph Stiglitz, the economics Nobelist and former chief economist for the World Bank, told me later he understood the protesters' frustration. 'This isn't how a market economy is meant to work. In fact, given everything that's happened,' he said, 'I'm surprised how moderate they are – surprised they're only "indignant" and not outraged.'

Theopi Skarlatos – Greece Has Gone Nasty – 08/11/12

As recession bit in Greece, reports of increasing far-right activity followed. There were stories of police collusion with fascist groups, violent protests against the left and numerous attacks on immigrants. Support for the openly anti-Semitic, anti-migrant Golden Dawn party soared, and it entered the Greek parliament in 2012 with over 9 per cent of the national vote. Despite being linked to numerous assaults – one of a female MP on live television – its figureheads won fans across the country and beyond. At its high point, the movement opened up offices in New York and Toronto. As more technocratic Greek MPs struggled to save the country's economy and its membership of the EU by pushing through tough austerity measures, Theopi Skarlatos saw a social crisis fast catching up with the financial one.

I'm standing in the square of Agios Panteleimonas, reminiscing about the Greece I used to know as a six-year-old girl. I remember playing in a similar square next to a decorated church, buying chocolates from the kiosk, skipping with my friends. Old Greek *giagiade*s (grandmothers) looking on. Somehow, back then, along with the summer pollen, the breeze carried an intense feeling that you lived in the best place on earth.

But now, off Acharnon Street, migrant children play ball games on a floor scrawled with graffiti saying 'Greece is for Greeks'. On the corner, plain-clothed policemen handcuff Somalis without papers and a one-legged man limps past on crutches, begging for money. The stone walls beneath the church are stained with the words *Chrisi Avgi* – Golden Dawn – the far-right party whose ideas are gaining popularity here at an extraordinary pace. Up the road, some more graffiti reads 'Down with the fascists!'

It seems the Greece of my childhood is nowhere to be seen. But people are fighting to try to reclaim it. They have to, they say, for the government has no real plan. Ahead lies only a sea of endless debt.

Another set of spending cuts and tax measures has been announced – 13.5 billion euros' worth. The prime minister is quick to point out the chaos that would ensue should they not be passed promptly by parliament. Rarely these days does he talk about the chaos that's already here.

Debts are rising; the economy's shrinking. There's high unemployment. An increase in drug use, depression, prostitution. People have been filmed foraging for food in bins after dark and schoolchildren are reported to have collapsed from malnutrition. A long way from the pretty pictures on dusty postcards at stalls beneath the Acropolis.

Golden Dawn, with its eighteen seats in parliament, gains voters by talking of a Golden Greece that once was. If it were in power, the party says there would be no more orders from Frau Merkel.

Immigrants ('invaders', as it likes to call them) would be sent home. 'Greece is for Greeks.' Members see it as their right to take law and order into their own hands. They have been accused of beating up, even stabbing immigrants, trashing their market stalls, and of stripping and humiliating women.

When we visit Golden Dawn MP Ilias Panagiotaros at his shop, staff are busy printing T-shirts emblazoned with swastika-like symbols. He tells me the party is preparing for 'a civil war' with 'everybody'. 'It's not nice,' he says matter-of-factly, 'but someone's got to do the hard stuff.'

The anti-fascists take to their motorbikes to protest. But some surveys suggest half the police force support Golden Dawn; they get little sympathy from the authorities. Some protesters tell us they were imprisoned and tortured by officers who openly praised Stalin and Hitler. The police deny these allegations, but forensic evidence later confirmed that some protesters did suffer 'grievous bodily harm by a sharp object'. The public-order minister has ordered an inquiry.

Before I leave, I go to see the director of the controversial play *Corpus Christi*, which depicts Jesus and the Apostles as gay men. He is Golden Dawn's latest target. On opening night, arm in arm with nuns and priests, party supporters manage to shut down the theatre. They glue the door locks shut and throw rocks at ticket holders.

The director cannot control his emotion when he tells me that his parents receive daily phone calls telling them their son will soon be delivered to them in tiny pieces. Three days ago, he cancelled all further performances.

Freedom of speech here is under pressure – as is freedom of the press. One journalist has been arrested for publishing the names of 2,000 well-known Greeks with Swiss bank accounts. Two more have been suspended from state TV after accusing the government of censorship. It has fallen to the international media to portray as true a picture as possible of what is happening in the country that once gave birth to democracy, but which is now struggling to show it is still in possession of it.

I no longer feel sad to be leaving a place that usually manages to hold on to my heart every time I visit. This time, I am sad about what I'm leaving behind. I am sceptical about whether I'll ever see that Greece again. The Greece that my six-year-old self, playing in the square, has kept alive in my memory for so long.

Emma Jane Kirby – Portugal – Families Split Again – 06/07/13

One of the bleakest aspects of the financial crisis was how much it over-shadowed the lives and aspirations of young people. In some corners of the EU – Portugal, Italy, Greece and Spain – youth unemployment had been high even before the recession began, and many of their young adults had left for work or study in richer member states. But as the crisis spread, opportunities close to home dried up, and many had to look much further afield. In Portugal, Emma Jane Kirby saw many telling signs of desperation – however stoically borne.

Somewhere on the line between Porto and Lisbon, my train slowed and pulled into another station. I glanced up briefly from my laptop, yawned and rubbed my eyes, and when I focused again I saw them, standing on the platform clinging to one another. The smaller man must have been pushing seventy: white-haired, wearing a faded candy-striped short-sleeved shirt, and his face, half buried on the taller man's shoulder, was screwed up in grief. From the way the taller man's back pulsed with irregular, jagged shudders, I could tell he was also crying. They were hugging so fiercely that, watching them, I found myself gasping for breath too.

Beside the taller man there was a battered giant suitcase and a dented, metallic trunk. As they pulled apart, I saw the same high forehead and finely chiselled nose on each face and it became clear they were father and son. And that this perhaps was their final goodbye.

I don't know them, but I bet I know their story. How the tall son, weary of rejection letters, panicked about the rent arrears, had gone to his father and said, 'Dad, there's nothing else for it – I don't want to leave you but I'm moving to Angola to find a job.' It might have been Mozambique, it might have been Brazil, but Portuguese professionals are now leaving their country in their thousands to beg their former colonies for work.

'Demand went up by 60 per cent last year,' Marcio at the Porto visa agency told me. 'In the first five months of this year, I've processed 45 per cent more visas on top of that. Engineers, architects, teachers – they're all leaving.' He sighed and twisted his pencil around his fingers. 'I see so many tears in this office,' he said. 'You know, when the visa comes through and they have it in their hand, that's when it hits them – the fact that they're leaving everyone behind.'

In a nearby suburb, thirty-three-year-old Flor, an architect, took me around her dad's pristine civil engineering offices, with their polished wooden desks and orderly in-trays. I noticed how quiet it was and asked her if everyone was at lunch. 'Lunch?' she said incredulously. 'Everyone's been laid off.'

We stood under a pencil sketch of her mum and dad on their wedding day: her father serious and suited, her mother, confidently beautiful, her hair piled into a Sixties-style beehive. Flor had been unemployed for nearly three years and her eyes were bruised purple with lack of sleep. In the last twelve months she'd been offered six jobs – five without any salary at all and the sixth with a token wage of 150 euros a month. Brazil, with its construction boom, seemed her only chance of finding work – but her mum, desperate not to lose her only daughter, begged and pleaded with her not to go. 'My mum died three weeks ago,' she whispered. 'So I know now it's time.'

Flor will not only be looking for a job for herself, but also for her elder brother Pedro. Her widowed father will not be coming with them. 'I'm so angry,' she says bitterly. 'I don't want to leave my dad. I don't want to work in a Third World country; Brazil actually frightens me with its level of violence, and I'll be completely alone. But what can I do?'

She looked up tearfully at the portrait of her mum and I understood at once that she was talking to her, not me.

This reluctant exodus to the former colonies has caused a massive social fracture across Portugal, leaving fissures and cracks in many tightly knit family units and fragmenting, perhaps forever, the relationship between parents and children, brothers and sisters.

Like thirty-year-old Claudia, an unemployed civil engineer and her elder sister Paula, who spent much of the time I interviewed them with their eyes locked on one another's faces, silent and motionless as if they'd been frozen in a photograph.

Claudia, in the footsteps of another sibling, had just applied for her visa to Brazil – she'll apply for Mozambique if it doesn't work out. With green-framed glasses and long dark hair, she looked intelligent and capable, but I could see that from where Paula was sitting, she was just her kid sister.

'I want everyone to get used to the idea that I'm going,' Claudia smiled bravely, 'so last week I told my little nephew. He looked up from the TV and he said to me, with his eyes full of tears, "You too? Soon it will only be me left in Portugal."'

'Well, at least we can Facebook,' she smiled reassuringly at her sister.

Paula smiled back but her chin was trembling. 'You can't hug on Facebook,' she replied.

Migrant Trail

Emma Jane Kirby – Italy – Just Arrived – 9/10/14

Just as more and more European citizens contemplated emigration due to the crisis, they were also confronted with more visible levels of immigration to the European Union – whether legal or illegal, refugee or 'economic migrant', asylum seeker or 'sans papiers'. Hundreds of deaths in the Mediterranean during the summer of 2014 drove the European Union to announce a new maritime operation called 'Triton', aimed at saving migrants on the sea route. Italy's government had been demanding that Brussels take full control of the problem, and the expense of policing it. The EU promised to send two planes and three patrol vessels but stressed the idea was to support rather than replace the efforts already being made. Emma Jane Kirby went to the Sicilian port town of Syracusa to see how Italy was dealing with the influx of people.

The sick are led off the cargo ship first, followed by the pregnant women. I count five who look to be at full term – the ship is three hours late docking because one has already gone into labour on board and has had to be airlifted to hospital.

Most of the migrants are African: Nigerian, Sudanese, and Eritrean. They all file down the gangplank onto the quayside with the same impassive, bewildered gaze, squinting into the floodlights at the swarms of police, charity workers and doctors who surround the boat, documenting and dividing them into groups. The barefoot are handed fluorescent flip-flops and garish plastic sandals. The migrants hold these in front of them as if they were visa or identity papers, some proof of belonging.

A chubby boy of about ten, wearing a chequered bomber jacket, suddenly jumps in front of our camera to give us a double thumbs up, jiggling with excitement about being on the telly. His head-scarfed mother puts her arm around him proudly and beams into the lens, talking quickly in Arabic, indicating that the two other teenage children behind her are also hers. I snatch fragments of her story: her husband has been killed; she's a Christian. I think she's Syrian; she says she's come to Europe to give her children a future. She's gesticulating, anxious to make me understand the details, but a policeman shouts angrily at us and herds her away into a holding area.

This year alone, 130,000 migrants have arrived on Italy's shores, so it's fallen to Rome to foot the bill for the search-and-rescue missions. At the coordination centre, a giant electronic map is marked with green dots indicating interceptions. The incidents are now so frequent that off the coast of Libya, there's barely a chink of blue sea visible; the Mediterranean has been permanently stained emerald.

Syracusa's mayor has just been informed of the 200 new arrivals on the cargo ship. 'It's like a tap, that you can't turn off,' he tells me, shaking his head wearily. 'Every day there are more.' We walk together along the seafront and he bangs the railings with his hand. 'Am I wrong,' he asks rhetorically, 'or is this frontier Europe's frontier, as well as Italy's? So isn't this Europe's problem, not just ours?'

The Italian government says it's sinking under the exorbitant cost of saving lives at sea and is calling upon the European Union's border agency, Frontex, to take over. Frontex's real remit is to protect Europe's frontiers and to detect who is crossing them illegally. Ironically, it's no longer repelling illegal immigrants but is on a mission to rescue them.

Eight hundred metres above Syracusa, in a small plane on loan to Frontex from Spain (the agency has no operational equipment of its own), the tanned young pilot Captain Cicuda shows me the sophisticated radar and surveillance equipment the crew is using to detect boats in trouble on the sea below us. He's not allowed to fly over the North African coast where most of the shipwrecks have occurred, but must stay closer to Europe's shores. The captain and his plane are only here for a month; after that, Frontex must rely on the goodwill of another EU member state to replace them. Frontex has called urgent meetings to try to rally its partners: the mission is also running out of money.

We're forced to break off the interview because of sudden turbulence - we're flying into a storm and I need to fasten my seat belt. The rain thrashes against my window and the wind hurls the aircraft over and under the heavy clouds. Winter's on the way and the Mediterranean below us will only get more treacherous.

On the quayside the steady drip-feed of migrants from the cargo ship continues. This time all were saved, but next time . . . ? I spot the chubby Syrian boy, dancing impatiently as he waits with his family to be bussed to the already heaving asylum centres. His mother catches my eye and grins at me confidently, raising her hands to the heavens in a gesture implying 'thank God it's all over'.

It's only just beginning, I think. How long before she understands that she's washed up on the shores of a Europe that neither wants her, nor knows what to do with her?

Matthew Price – Hungary – Walking Alongside – 12/09/15

The 'migrant summer' of 2015 wasn't only – or even primarily – a matter of dangerous journeys by sea. There were also hundreds of thousands of people on the move over land, drawn by Germany's commitment to look after those fleeing war in Afghanistan, Syria and Iraq in particular. Chancellor Angela Merkel argued that there were 'extraordinary circumstances' which justified letting them in – and Germany did accept over a million applications for asylum during the year. Many were made by people who had taken exhausting and complicated routes into Europe, from Turkey to Greece, and then through the Balkans. Train stations piled up with crowds of families; bus convoys crawled along the motorways. And at the roadsides, thousands more trudged onwards on foot for day after day. Matthew Price spent some time walking beside them.

She was limping slightly. The pushchair was weighed down. Her handbag hung from the left handle, a bag of apples from the right.

Underneath was a blanket, stuffed below the seat. On the seat more belongings, all they'd been able to bring, and on top of the lot was her three-year-old, balanced precariously, fast asleep.

Her son walked alongside. He and his sister, neither more than six or seven, had dragged themselves along this, the hard shoulder of the motorway, without complaint or slowing for ten hours. But now, in the darkness of the Hungarian countryside, with Budapest's railway station more than twenty miles behind them, it was becoming too much.

The boy began to wobble slightly, then ground to a halt, putting his little hands on the pushchair to support himself. His head fell downwards and he cried out. His mother stopped pushing to comfort him. And then, realising they had to move onwards, yet there was no way of carrying her children, she too began to crumple, hands on her legs, slipping downwards. 'We are so tired,' she managed to say, before the tears came – before she was overwhelmed. Her daughter stood there, not knowing what to do. In the fast lane, the vehicles drove by.

Others passed. Some, also exhausted, sat or lay up against the crash barrier. A thousand or more people, strung out along the main road to the Austrian border. The men marching up ahead – single and in the majority – had set too fast a pace for the families. The group was drifting apart. Occasionally a police van drove past, warning people of the dangers, asking them to move from the carriageway.

The woman continued with her daughter. A stranger pushed the pushchair. Another carried her son on his shoulders. The boy held on tightly, sitting upright for a few moments, but then tiredness overcame him, and his head slumped forwards as he fell into a deep sleep while they walked onwards. The boy woke, startled, an hour later, crying for his mother in the darkness. And there they stopped, having finally caught up with the group. Blankets were shaken out; the boy clutched a small teddy bear and lay down. And the cars kept passing.

That was the last I saw of them. I assume they are now in Austria or possibly Germany. A few hours later the Hungarian government sent a hundred buses to pick up those stranded along the motorway. Left on the Hungarian side of the border, they trudged across into Austria, through the pouring rain.

I asked one family how long they thought Syria's war will continue. For many years, came the reply. That will be the case for Libya, Afghanistan, and beyond. Refugees from these conflicts will need a safe haven for a long while yet. And what of those not directly affected by war, but whose countries are still affected by the fallout. Or those looking for better prospects – better jobs, better schools for their children.

This is a human story, a political story, a story of war – and a story of mistrust. On social media many people have responded positively to my reporting. Others have been appalled at what they see as my 'bias' – my support for these people. They've responded with crude, visceral anger. Don't I know these people are coming to rape our women, to chop off our heads? ISIS is among them, they claim. It's a terror march.

This is a time of immense importance for all of Western Europe. The human migration will continue, even if winter slows it temporarily. Without careful political handling, the divisions within our societies over what to do will grow.

Days later, at Munich railway station in Germany, a group of 200 young men waited to board a train north. Among them a young man, clean despite days on the road, good-looking, with a clear smile. He was a musician from Raqqa in Syria, the town taken over by ISIS. 'They said they would surely kill me, would kill all musicians,' he said. He'd had no choice but to go. Was he a Muslim? 'Yes, but I hate ISIS,' he said calmly. 'I love my country. It is a paradise. And one day I will return.'

Nick Thorpe – Slovenia – Trapped – 24/10/15

As in any emergency, the correspondents covering the scenes of mass movement often found themselves wondering how much they could – or even should – help the people whose journeys they were reporting on.

Nick Thorpe found that in some moments, information was a far more precious commodity than rest, food or water; what many migrants wanted more than anything was to know where to go next.

In a moment of weakness, I lent my phone to a tall, sad-faced man. It was early in the morning. The refugee camp on the outskirts of Lendava was full of new arrivals. So there we stood, in our little puddle of technology, sprouting phones and cameras, modems and cables, suddenly surrounded by people desperate to trace loved ones they had lost in the stampede in the rain through Croatia. The batteries on their own phones were dead, their SIM cards from Greece, Serbia or Syria rendered useless.

Everyone seemed to have lost someone. Delshan her father. One man his wife and three children. He was crying all the time. 'Why didn't you stay close to them?' another man wants to know. 'The Croatian police only let women and children onto the bus,' he sobs. Many accuse the Croats of brutality. Of herding them like animals across their country, the last one before Slovenia. Even of beating them. 'Please, your phone? A short call to Kabul? To Damascus? To Austria?'

My nose starts streaming in the cold. A small boy notices, and hands me his little packet of tissues. I take one, hesitantly, from his five; 20 per cent, perhaps, of his worldly possessions.

The grape harvest is long gone, but the Balkan grapevine the migrants use to find out where they can go, what they might expect, and even who is bombing their country now, buzzes with news and rumours. Europe is having second thoughts about accepting more migrants, they realise. They are in a mad rush, before the draw-bridges are drawn up in front of the gates of all the castles. Before the archers appear on the battlements.

The refugees don't even stop to sleep, unless they are delayed by borders or bureaucracy. The razor-wire fence built by the Hungarian government along its borders with Serbia and Croatia has become a yardstick by which other governments of the region measure their own response. An off-the-cuff answer by a Slovene minister to a journalist's question is suddenly a global headline:

'Slovenia considering building physical barrier'. The Slovene parliament's vote to bring in the army provokes a similar interpretation. Visions of men in NATO kit with big guns, keeping out hordes of bedraggled migrants. Or will they just shepherd them onto buses?

All day long, our phones ring. Relatives returning calls. Someone's mother from Kabul. A string of country codes I don't recognise. A torrent of Arabic, Farsi or Pashtu. Occasional fragments of English. We've already left that camp. Your son is OK! In Slovenia? Which camp? But what do place names like Sredisce ob Dravi, Brezice and Dolga Vas mean to people in Kunduz, in Kandahar or in the ruined suburbs of Aleppo? One man from a tiny Afghan linguistic minority, with a magnificent nose, who has lost his pregnant wife insists on speaking to me in the quick-fire Greek he learnt working for a NATO battalion.

I answer a call from Austria. It's a man from the Catholic charity Caritas, trying to help Hassan, an Afghan refugee who's now in the Austrian village of Neudorfl, find his mother, his aunt and his sisters. One of them used my phone twelve hours earlier. My colleague Orsi goes back to the camp with the photograph which Hassan texts to her. It's already dark. There are so many Afghan families, leaving and arriving. Everyone shakes their heads. Then Orsi hears running feet, and is pulled excitedly into a huddle of women by a big tent. The phone is passed round.

Hassan's mother, wrapped in a huge blanket, starts hugging her. An explosion of joy in the chill Slovene night. Orsi takes a picture of the family to send to Hassan. There's sadness as it turns out that Sediqe, his fifteen-year-old sister, is still missing. Meanwhile, Delshan from Syria, witnessing one family reunification, can't hold back the tears. There's still no sign of her father.

Earlier in the day, the boys played football with car tyres for goalposts. A high-scoring game – the goalkeepers understandably reluctant to dive on the hard tarmac, after risking life and limb so many times on their long journeys. This whole refugee crisis seems to me like a football match. Rich nations of Europe 2 – Wretched of the Earth 3. But how long will the match last? Can we play on their team, and they on ours? Who's the referee? And what might victory mean, for either side?

Jenny Hill – Germany – Mama Merkel's Backlash – 17/09/16

For the most part, Germany's commitment to housing and helping asylum seekers held good, and without the mass social unrest which some doomsayers had predicted. But there was still grave public disquiet about a spate of sexual assaults during New Year's Eve 2015 celebrations in Cologne, apparently committed by young men of Middle Eastern or Afghan origin. Right-wing groups made political hay from fears about migrants and crime, despite little statistical evidence that migrant men were any more likely to commit crimes than Germans of the same age and income. Jenny Hill asked whether there would be consequences at the ballot box for the Chancellor or her party, the Christian Democratic Union.

The dog – a tufty brown scrap of a creature – was not happy. Its owner carried it down the groaning pontoon onto the waiting boat, its legs outstretched and rigid with protest at leaving dry land. The captain glanced at it, glared at his other passengers and turned to the tiller. The engines spluttered and off we cast into the bright, choppy waters of Hamburg's famous port.

For centuries the tides have carried wealth and people in and out of this city. It's used to change. Even so, for Hamburg – for Germany – it's been a year like no other. Thirty thousand people have arrived – and stayed – here during the last twelve months, hoping to be granted asylum.

Take Abdul, a young man, a welder by trade, from Syria. When I met him, he was hard at work, blue sparks flying across the workbench reflected in the visor over his face. He lifted it, revealing startlingly green eyes and a big grin. He's already found friends here, a new home and a job. 'It's like building a house,' he told me, excitedly. 'You have to put down the first stone. This is the first stone.'

Abdul's new boss, a great kindly bear of a man, came over to check Abdul's work and slapped his shoulder with an approving paw. Holger Erren makes parts for the airline industry. Business is good, he smiled. He needed a new worker and thought: why not employ a refugee?

'After all,' he said, 'it's nearly impossible to find young Germans who could do these jobs. Abdul was in a camp for one year and he's really keen to work. That's something for our company. He motivates the other employees. Already they've asked him: "Why do you work so much? Why do you stay so long? Normally we stop at four!"'

This is exactly how Angela Merkel would like the refugee crisis to end: well-integrated migrants helping to mitigate the effects of an ageing population and bringing their skills to German business. Her government aims to get half of the new arrivals into work within five years. But it's easier said than done.

In a plain office, tucked away at the top of a sprawling Hamburg shopping mall, sits the woman responsible for integrating the city's refugees. Perhaps it's no surprise that Senator Melanie Leonhardt looks tired. A third of the new arrivals here still live in emergency accommodation, like sports halls. And it's hard to recruit enough language instructors to teach them German. 'We're not naive,' she says. 'We don't think people arrive here and fit in immediately. It's hard work, but we have no alternative. Either we turn this into a new chance for our society and a success, or we fail.

'It's not just that we *can* do it,' she adds, a little fiercely. 'We *have to* do it.' She's referring to Angela Merkel's over-used, and now rather derided phrase – *Wir schaffen es*: We can do it.

It makes me wonder. I've heard the same sentiment from dozens of other town and city officials over the last few months. The number of refugees coming into the country has fallen significantly – largely due to the closure earlier this year of the so-called Balkans route. The people managing housing, integration classes and asylum applications say they can now – just about – cope. They can see a way to make the process work. But it will take time.

That's time Angela Merkel doesn't have. German voters

are running out of patience. Just a few weeks ago the electorate in her own home state delivered a humiliating verdict at the regional ballot box, knocking her conservatives into third place behind the anti-Islam, anti-migrant party, Alternative für Deutschland.

I recently listened to one of AFD's founders speak at a hustings. Islam, he declared, is 'a religion of blood which has no place in Germany'. The recent arrivals are 'culturally behind us', he said. 'They can't or don't want to integrate.'

For the first time, Mrs Merkel's political future looks genuinely shaky. One of her own, very loyal, conservative MPs told me sadly that the chancellor now personifies the refugee crisis. That she's perceived to have lost control, to have opened the country's doors to terrorists as well as refugees.

Mrs Merkel has urged Germans not to fear superficial change, and promised that the country's fundamental values won't alter. But as the tourist boat docks and I walk back along Hamburg's harbour front, admiring the elegant tall ships that have recently put in, I wonder how many Germans believe her – and whether there's much more she can say to persuade them.

Ben Zand – Libya – The Deceived – 23/09/17

International efforts to control migration have often been centred on enforcement: barriers and borders, more intensive policing, more deportations of those without the correct papers. Since the migration surge of 2014–15, the European Union has increasingly opted to try and subcontract this stage to other countries further 'upstream' – like Turkey, Libya and Niger – financing them to find and detain people already en route. And there have been attempts to provide alternatives to illegal migration, by subsidising programmes providing jobs, business advice or grants in migrants' home countries. But after following the migrant trail from West Africa, over the Sahara and the Mediterranean to Europe, Ben Zand was struck by what seemed to be missing: better education about the danger and costs of the journey itself.

Lies, lies and more lies. The migrant trail is plagued with misinformation. That's what I think to myself as I sit, exhausted and splattered with water, on a speedboat heading to Tripoli. The road is too dangerous, so boat is the best option.

I've been up all night with the Libyan coastguard. They're pretty terrifying: most have scars torn along their faces and the leader is missing three fingers – lost to a bomb blast years ago. In just six hours at sea, we find 600 people – men, women, and children. They're piled onto four shoddy boats made out of what look like crates, which don't seem at all as though they're made to be in the sea. We approach, with the coastguard armed with RPGs and machine guns.

'We'll get to Europe and they will welcome us – we'll be given jobs, and we will get money,' a man from Niger tells me.

Back on shore, a man called Ghassan – from Iraq, fully suited, in his sixties and holding a briefcase – beckons me over.

'Sir, can I just explain my situation? This is all a misunderstanding,' he says as I walk over to him. 'I'm an English teacher. I've been in Libya for fifteen years and I'm just making the trip to Europe because I have a heart problem. My medical records are in this briefcase.'

He gestures: 'I paid a smuggler £2,000, and he told me that I just need to get a hotel for the night, and then in the morning I would get the ferry to Italy.'

'The ferry?' I ask, confused.

'Yes, he never said it'd be a boat like this. He said it'd be a nice boat.'

Ghassan wants to know what's going to happen next. 'How do I get to Europe from here?' he asks me optimistically.

I stutter, not knowing how to reply. Does he really not know what lies ahead? I've spent quite a bit of time reporting from Libyan detention centres. Black holes where migrants go and many never return. They are not places for anyone, never mind a man in his sixties with a heart condition.

'I don't think you go to Europe from here,' I reply, asking myself: do you really want to be the one who tells him? I continue: 'The coastguard says you're going to a detention centre.'

'And then I will go to Europe?' Ghassan asks pleadingly, looking for the light at the end of this crumbling tunnel.

'I don't think so, Ghassan. It's most likely you'll be sent back to Iraq.'

'I can't go back!' he screams. 'I don't have family there. Please, it's a misunderstanding.'

He begins to cry. And so, nearly, do I. How did he not know these risks?

It's not just him. As I travel along Africa's migrant trail I hear things I just can't believe. Was it naivety or desperation?

In a smuggler's compound in Niger a man tells me what he knows about Libya: 'They love nature; they love people.' His friend adds: 'We will make money there'.

'Do you not know it's dangerous?' I ask.

'No,' they reply.

I hear stories of women being raped, beaten, used as slaves – men being held captive and forced to work. But when I put this to people along the migrant route, they don't quite believe me.

'I've heard horrible stories,' I explain to two young, well-dressed women who are on their way to Europe to work in the sex trade. 'I've been to Niger where people are dying in the desert; I've been to Libya where—'

The girls cut me off. 'In life there are always risks,' they tell me. 'We will be amongst the successful ones.'

I meet a group of Nigerian men who were abandoned in the desert. Most of them have no shirts or shoes on. 'Did you know the risks before you left?' I find myself asking again.

'No, we didn't. We heard some things, but we also have a lot of friends who have made it. Successful people post pictures online next to fancy houses, next to big cars – I want that life. Europe is sweet; back home it is not.'

In Nigeria, I finally meet a smuggler. 'Over two thousand people have died in the Mediterranean this year so far,' I say to him. 'Do you tell your customers about the risks?'

'No,' he replies. 'I have to think of my own family. If I told them, they wouldn't come and I wouldn't make any money. They search for money and so do I. My country is broken.'

I sit there, shocked silent, and realise just how strong the European dream is. Until life improves at home, it'll remain an easy dream for the smugglers to sell.

Part II: AFRICA

Africa

Jo Fidgen – Zambia – Bad Husbands – 03/04/2010

At the start of the decade nearly a million Zambian citizens were HIV positive and there were more than 95,000 deaths a year from AIDS and related illnesses. Although antiretroviral treatment was already slowing the rate of new transmissions, the epidemic's impact on this country of around fifteen million people was immense. The costs and consequences of caring for so many patients rippled far beyond the health system into most families. And in Zambia HIV prevalence was skewed by gender, with 60 per cent of those affected being women and girls. Jo Fidgen traced how sexism and a public health emergency exacerbated each other.

I'm relaxing with a cup of coffee when my good friend Margaret smacks me round the head. She doesn't use her fist – she's not the violent type – but what she tells me has the power of a heavy-weight's punch.

I'm always soothed by her company: she glides through life, despite a seemingly endless supply of problems. In the time that I've known her, she's lost one room of her four-room house to floods; a distant relative has left a four-year-old boy with her and disappeared; and her husband, who works as a mechanic but never gets paid, has been constantly ailing – with malaria, cholera, diabetes.

Then, out of the blue: 'Did I tell you my husband is on antiretrovirals?'

She reads from my face that she had not.

'He lied to me when I married him,' she continues. 'He didn't tell me when I was pregnant with our first child. Not even with our second.'

She is serene, apparently above – or maybe beyond – anger. I, on the other hand, am floored.

'He only told me when his eleven-year-old daughter from his first marriage came to stay,' Margaret adds. 'She was dying.'

She pauses, and stares into the middle distance. 'I shan't get married again when he goes. It's better not to be a wife.'

I really shouldn't have been surprised. Official figures suggest that the greatest proportion of new HIV infections here occur in married, middle-aged women. Part of the problem is that wives can't insist on using a condom: to suggest it would be to imply you are being unfaithful, or suspect your partner is.

But carrying on two or more relationships at the same time is seen as acceptable for men – as I heard from no less an authority than a doctor at the National AIDS Council, a government agency. 'Define extramarital sex,' he challenges me. 'After all, a man may have more than one long-term partner. They are just like wives. Are you saying he should be treating them as casual girlfriends and using condoms?'

'But it says here,' I reply, waving a glossy document prepared by his very organisation, 'that multiple concurrent partnerships are the main reason HIV is spreading in Zambia. How can a man be sure his girlfriends aren't sleeping with someone else?'

He bristles. 'Why are you so against us, against the government?'

I am baffled by this, but he won't elaborate.

Later that day, I stand in the kitchen of a health campaigner, trying to make sense of the discussion I'd just had. She arches an eyebrow and invites her cook, Hope, to tell me her story. 'I'm HIV positive,' Hope says. 'My husband got it from one of his girlfriends and gave it to me. He's never apologised.'

'It's the biggest risk a woman can take,' adds her employer, 'getting married.'

A wife has a strange place in Zambian society. Marriage is a passport to respectability. 'Single women are prostitutes,' a traditional counsellor once told me, advising me to hurry up and get on with it before no man would have me. According to custom, wives must submit to their husbands. I went to one wedding where the speaker solemnly told the bride that she must never contradict her man, and should accept that he was entitled to other women. The newly wed was a city girl with a prestigious job. I wondered what she made of the advice.

Very few women will give up on a marriage because to do so would be to lose their position in society. Perhaps that's why Margaret didn't opt for divorce. She was lucky: neither she nor her children had been infected. I don't know what shocked me most – that her husband, a likeable man, had thought it acceptable to gamble with the lives of those he loved, or that she had forgiven him for doing so.

Who knows why they took those decisions, but I suspect the power of marriage and the powerlessness of women played their part. Similar forces may have been at work in yet another disconcerting conversation, this time with a very forceful, very successful government official. I'm packing up my bag at the end of our meeting when suddenly she asks if I am married. 'Husbands are bad,' she warns me. 'Mine died of AIDS. I'm carrying the virus.' To my astonishment, in the very next sentence she urges me to tie the knot.

It sounds like a contradiction to me, but something more complex is going on. Only through marriage can a woman claim a place in society. And if the price is to put her health on the line, so be it. HIV has coupled itself with inequality, and you fight them together, or not at all.

Justin Rowlatt – Angola – Karaoke with Chinese – 02/10/10

A major story across Africa, as in Asia and Latin America, was the extraordinary reach and ambition of China. From vast infrastructure projects to ambitious small-time traders, China was willing to put a lot into the continent: scouring it for resources and opportunities, trying to match African needs with Chinese skills, goods and credit. Justin Rowlatt asked what was driving the young Chinese who were willing to venture in search of profit.

We'd done our best to put our new friends at ease. We'd chosen a picturesque Chinese restaurant on the beachfront with dragon motifs and Chinese lanterns swinging in the breeze. We'd laid on a supply of chilled Tsingtao beer and the waiters were bringing out steaming plates of delicious Chinese food. Yet we still didn't seem to be able to get past a kind of anxious 'first date' formality.

Jet told me he'd set up an air-conditioning business here five years ago. He laughed and chatted nervously about the trials of life in Africa, and his ambitions for his firm. Betty, a twenty-two-year-old Chinese woman who's got various projects going here – including the local Chinese-language newspaper – buzzed among our guests, gossiping and giggling, but even her charms failed to put our guests at ease.

I was starting to fret that we'd never break the ice. Then our translator, Poppy, put on the karaoke machine. Just as she had promised, the atmosphere was transformed. In a moment, even the shyest of our guests were up and competing to choose a song.

To get things started Betty and I sang an unlikely rendition of 'I Will Always Love You'. Betty sang beautifully . . . I did not. But no one seemed to mind. Chinese karaoke appears to be more about taking part than tunefulness and our guests were now competing for the mic. At last, our little party was beginning to swing.

Foreign travel is still rare for Chinese people and Africa is considered a very exotic and dangerous destination. 'Many of my friends in China have never even heard of Angola,' Wang admitted later as he picked at a plate of sweet roast duck. Like Jet and Betty, he's an entrepreneur, here to run the African arm of his uncle's aluminium-window company. 'At first I found it frightening,' he confessed. 'You hear lots of stories of Chinese people being robbed by the locals.' So why did he stay, I wanted to know? 'There are great opportunities here,' he replied without hesitation.

You hear the same from Chinese entrepreneurs all over Angola. After twenty-seven years of civil war, the Angolan economy – once one of the strongest in Africa – was in ruins. 'Everything had been destroyed,' recalled Jet, remembering what it was like when he arrived here shortly after the war ended. 'There were no roads, railways, shops: nothing. Several Western companies were already here selling their products, but I knew I could import things cheaper from China.'

Betty nodded vigorously. She's only been here a couple of years, but also sees real potential in Angola. 'I'm doing much better here than if I'd stayed in China,' she said.

It's not just small businessmen and women who have come. Some of China's biggest construction firms are here building roads, railways, hospitals and vast housing complexes. Vera and Deng both work for state-owned construction companies but, like our other guests, they reckoned they'd be able to get on quicker in Angola than back in China. 'I earn twice as much as I would at home and I've got a better job,' explained Deng. 'Chinese companies can be very traditional and it can be hard to get promoted.'

Looking around, two facts stood out about the people at our table: how young they all were (the oldest was in his early forties) and how well-educated they were. Virtually all were graduates and most spoke excellent English. You might imagine young graduates like these would have a bright future in China, well on their way to becoming the country's new elite. But our guests – and tens of thousands of others like them – decided to take the risk of travelling in search of opportunity.

There are now an estimated 100,000 Chinese people in Angola, and around a million across Africa. Their presence reveals an essential truth about China which is often overlooked: in terms of per capita incomes, despite its dazzling economic growth, China is still a very poor country. It can be hard for even the best-educated Chinese to find a good job. The latest Education Ministry figures show more than a quarter of this year's 6.3 million Chinese university graduates are now unemployed.

By the end of the evening we were all a little worse for wear and Jet was back at the mic singing a mournful Chinese ballad. He'd told me he would like to return to China one day, but it was clear there'll be a few more karaoke nights here in Angola before Jet returns home.

Alastair Leithead – Gambia – Goodbye, Jammeh – 26/01/17

For twenty-two years, Yahya Jammeh ruled Gambia with an iron fist and a crushing security state. His eccentric public pronouncements, bizarre and ineffective herbal 'cures' for HIV, and intimidating entourage only amplified his oversize persona. But in 2017 Gambians overcame their fear and voted him out. After initial resistance, Jammeh was pressured by other West African heads of state, and a hastily arranged regional military force, to go into exile in Equatorial Guinea. Gambians who'd fled during the stifling Jammeh years began coming home with high hopes for the new administration of Adama Barrow. Alastair Leithead reported from the capital, Banjul, on the country they'd find.

After long, hot days, the sun sets quickly in the Gambia. Approaching the horizon, it almost rushes to bring the day to an end. But a president who'd had his twenty-two years in the sun appeared to be in no rush to follow suit, lingering on as a long and brutal period in this narrow strip of a country's history came to a close. Standing on the tarmac at Banjul airport, a gaggle of gathered journalists watched the sun slip away behind the small, white, private jet that stood waiting, the photographers frustrated that when former President Yahya Jammeh left the country, it wouldn't be in a good light. When dictators fall, that's often how it works out.

As hours ticked by, every little movement was interpreted. The heavy holdalls being loaded onto the plane: sacks of cash? Records of the terrible things that happened to many Gambians suspected of dissent? The long swathes of red carpet were rolled out – as they had been on the previous day, only to be rolled back up again. But when the military band arrived and started warming up, it appeared that the waiting was over. Neighbouring heads of state, and the heavily armed regional troops gathering on the border, had persuaded the man who'd lost the election and admitted defeat, only to change his mind and refuse to leave, to finally head over the horizon.

It was dark when the sirens blared and the motorcade swept

across the tarmac. In his flowing white robes, Jammeh stepped out of the car and onto a platform as the band stuck to ceremony, observing presidential protocol for the last time. From a crowd of aides and well-wishers he emerged onto the aircraft steps, kissing the Koran, waving it to the crowd, and disappearing from Gambia's life for the time being.

That's when the rumours started. Was $11 million really missing from the central bank? Did a Chadian cargo plane follow him, carrying a fleet of fancy cars and a mountain of luxury goods? Had he really booby-trapped State House? Had he negotiated himself immunity from prosecution?

A lot about life in the Gambia has changed extremely quickly, but the secrecy and suspicion linger. A lifetime of living under the threat of arrest, torture or 'disappearance' leaves its mark. 'We've always spoken like this,' the border policeman whispered as we arrived in the country, hiding his mouth behind the back of his hand. 'But now we are free,' he grinned, removing it. 'You're welcome, BBC!'

The T-shirts being sold on the streets shouting 'Gambia Has Decided' would have led to instant arrest just a week before. After voting for change in the election, Gambians now have a voice and they're ready to use it. The new president, Adama Barrow, not only needs to listen, he needs to act – to show the real leadership people are so unaccustomed to, and somehow turn a poor country around. He has a river of goodwill (swelled by expectation) running the length of this country, but it won't carry him smoothly forever. The eddies of coalition politics will catch up with him and swirl around a man who probably didn't expect to be president when an alliance of opposition parties plucked him from relative obscurity and dropped him into the rushing waters.

What of the missing millions? It's a story to offer the 39 per cent who voted for the former president. But it's not proven; either the coffers are intact, or the money is gone, depending on who in the new presidential team you listen to. Yes, the luxury cars are following him to exile in Equatorial Guinea – that was apparently part of the exit deal. Now the doves are pushing for truth and reconciliation, while the hawks want to make sure he never comes back.

There's an expression everyone knows here in the Gambia: 'the back way'. 'Where did that boy go?' you may ask of someone. 'The back way,' they reply. It means the migrant trail to Europe: east to Niger, north through Libya, across the Med and into Italy or Greece. Villages have emptied; the money-transfer business is booming thanks to remittances from those who make it. Former President Jammeh has now forged his own 'back way' to Equatorial Guinea, and many hope he'll stay there ... while the new leader of this river state has the light of a new day, to boost the economy, bring jobs and stem the flow of his people. But there will be rough waters ahead.

Gabriel Gatehouse – Liberia – The Ambulance Driver – 30/10/2014

The West African Ebola epidemic of 2014–2016 killed over 11,000 people – and the world waited nervously to see how much further afield the contagion might spread. Ebola is a terrifying disease: the symptoms are painful and its progress frighteningly fast. Any infected person's bodily fluids can be highly contagious. And perhaps worst of all: in a region where funeral rituals are extremely important and families gather together to wash, dress and bury their dead, corpses are likely to infect mourners. Medical staff trying to treat the sick and limit the spread of the epidemic were risking their own lives – and they knew it. Gabriel Gatehouse learned just how brave they had to be.

Laurene winces as she walks over. She's in pain.

We're in the yard in front of her house. It's at the end of a quiet sandy lane of bungalows and suburban houses. The sound of children playing basketball drifts from a nearby playground. She needs this refuge of calm in her downtime. Her job is one of the most dangerous – and distressing – in the world. As part of a three-person ambulance crew, Laurene travels around Monrovia picking up Ebola patients and bringing them to special clinics for treatment.

Laurene motions to her teenage son to bring her a chair so we can sit down while we talk. She smiles as I sit just a little further away than would normally be appropriate. She understands. By now, she's used to it. She is in direct contact with Ebola every day: people want to keep their distance.

Here, at home, Laurene looks tired, almost vulnerable, behind the eyes. But there's no mistaking it. This is a tough lady. Tough in an 'I've seen it all and can take whatever the job throws at me' kind of way.

When Laurene goes into Ebola-infected homes, she can look terrifying, and she knows it. Any skin must be covered. Behind the goggles and the mask, it's impossible to read her facial expressions. Taped into her yellow gloves and white biochemical hazard suit, hood up, Laurene looks like a creature out of a sci-fi horror film.

But she's a nurse. When she goes into an infected home, she is accompanied by a driver and a 'hygienist' – a young man by the name of Zach. In their outfits, the only thing distinguishing Zach from the other two is the vat of chlorine strapped to his back from which he is continually spraying.

Laurene explains how she injured her back. The team was called to the home of a woman suffering from the advanced stages of the virus. Laurene went in first. The patient was lying naked on a mattress stained with blood and vomit. It was a highly dangerous environment.

The patient was in a bad way, too weak to stand or get dressed by herself. Laurene covered the woman with a wrap and went behind her, to pick up her emaciated body under the arms. Suddenly, and with a ferocity Laurene had not expected, the woman turned on her, pushing, shoving and shouting abuse. Laurene fell over backward, onto the corner of the metal stretcher. Her first thought was her suit – a rip could have had fatal consequences. She left the patient where she was and went out to check for tears. Luckily, the suit was intact, so Laurene should be safe from the virus. But she's still hurt her back.

Laurene says she doesn't blame the woman who attacked her.

She knows and understands the fear. 'I've had sick patients jump out of my moving ambulance before,' she says.

Residents of Monrovia occasionally catch sight of a pickup truck laden with watertight orange bags, bumping over the city's potholed roads. These are the bodies of the dead, being driven to an anonymous site for cremation. Many believe this is the inevitable fate of anyone who enters an Ebola treatment centre.

That's not true. If you catch the virus, your best hope is to seek treatment and seek it fast. You can survive Ebola – in fact many people have, and continue to do so. But the number of those who live is still far lower than the number of those who die. Often, the last a relative will see of a loved one is them being stretchered off by what look like creatures in space suits, spraying chemicals as they go.

I ask Laurene whether she was upset by the sight of the sick woman on the mattress. 'I've seen worse,' she says, brushing the question aside. Laurene has collected hundreds of Ebola patients since the outbreak started. She has no idea how many of them have died, how many have survived. There will be many more to come. If she is to continue her work, she must protect her mind as well as her body. Beneath the goggles, she wears an emotional mask too. An excess of empathy is a luxury she simply cannot afford.

Is she ever scared, I ask? 'I am not afraid,' she answers. 'If you have too much fear,' she says, pointing at her head, 'you will get affected.'

Lorraine Mallinder – Guinea-Bissau – The Game Remains the Same – 14/04/18

Guinea-Bissau is a small, poor, underdeveloped state on Africa's Atlantic coast, with a long history of political turmoil, strings of military coups and more than two-thirds of its people living below the poverty line. And just far enough north to be a convenient stop on international drug-trafficking routes, where cargo worth millions can be shipped with relative impunity. Guinea-Bissau was described by some United

Nations sources as being at high risk of becoming a 'narco state', where narcotics profits completely permeated the country's economy. Lorraine Mallinder saw both drugs and drug money were circulating widely: a story of unstoppable globalisation, and the criminal networks binding continents together.

'*Um quilo de cocaína.*' A kilo of cocaine. I've just been made this startling offer on the streets of Bissau, capital of Guinea-Bissau.

Abu – in his early thirties, casually dressed, evidently works out – used to be a mule. He'd carry capsules of cocaine in his stomach from Brazil to Bissau for €750 a trip. But now he's moved on to bigger things, coordinating trafficking from Bissau to Lisbon. Don't be fooled by his battered white van – 'a poor man's car', as he puts it. 'Oh, I have some very nice cars,' he says. 'But I don't want to stand out.'

Now he's trying to sell me a kilo of cocaine for €20,000, which, he says, will fetch €50,000 on the streets in Britain. Or €80,000 if I sell it in Switzerland.

'Erm, not really my thing,' I reply with nervous laughter.

'Yeah, well, it's a dangerous business,' he says, looking slightly irked.

Earlier, while he'd recounted his experiences as a lowly mule, I'd noticed the upwardly mobile trafficker was growing antsy. 'I'm beginning to wonder if you're with Interpol,' he'd said.

Back in the 2000s Guinea-Bissau fell prey to Latin American cocaine cartels looking for new routes to shift their product to Western Europe. It was an ideal transit point, its unpoliced coast dotted with dozens of mostly uninhabited islands, its politics chronically unstable. The past twenty years of its history have included a civil war, three coups, and the assassinations of a sitting president and an army chief in what was thought to be a tit-for-tat double murder linked to drug profits. Bear in mind, no elected president has ever managed to see out a full term here.

Back during the narco boom, around 2008, the UN reported that around fifty tonnes of Latin American cocaine were being shifted each year through West Africa, much of it landing by boat or plane in lawless Guinea-Bissau. In 2013, America got involved, with undercover agents posing as members of the Colombian rebel group FARC, netting a former navy chief in a cocaine-for-arms sting worthy of a Hollywood blockbuster. The high-profile operation is thought to have spooked drug lords, with their goods reportedly diverted through neighbouring countries.

Yet it doesn't look that way on the streets of Bissau. One source lets me in on a Facebook Messenger conversation to show me just how easy it is to order a few kilos to be dropped off in the city, an exchange that takes about two minutes. 'I know so many people who are in it. They don't work, but suddenly they have new cars,' he says. Later we see one such character in the poor neighbourhood of Bairro Militar in a shiny new Nissan Micra – apparently a highly desirable vehicle in a land where dodging potholes is a national sport.

I visit the judicial police's narcotics lab, which is partly funded by the European Union, and am shown a bag of cocaine-filled capsules. Last year, the judicial police seized 8.5 kilos of the stuff, all carried by mules. Their director estimates that less than a tonne of cocaine now flows through the country each year. But, as he himself concedes, with no radars for detecting air traffic and no X-ray scanners to check shipping containers, it's hard to be sure . . .

Outside his office, there's a picket line. It turns out the judicial police themselves have been on strike for a week over pay and conditions. I recall hearing allegations that its poorly paid agents on the islands are easy prey for drug lords offering cash bribes.

I meet a high-ranking UN official for coffee. He estimates that thirty tonnes are still moving through the country every year. Big cargos come by plane, he says, some transported up the Atlantic by international fishing boats – legal and illegal – and some brought to land on small vessels by local fishermen, where it's moved by the military over the borders. The players and alliances are constantly shifting, with groups like the Islamic State in the Greater Sahara now said to be negotiating directly with Latin American cartels. 'It's still a major hub. The actors change; the game remains the same,' he says. 'Cocaine is so plentiful here, it's like a currency,' he laughs. 'I owe you 100k, I give you 250k's worth and you get rid of it.'

A couple of days later I bump into Abu again, driving his 'poor man's car' through Bairro Militar. I tell him about the judicial police's estimate of one tonne of drug transit a year. 'They would say that!' he scoffs. 'Of course the planes and boats are still coming in. They can't be detected.'

'Are you *sure* you're not interested?' he asks, still dangling his offer. Guinea-Bissau, it seems, is still very much under the influence.

Katerina Vittozzi – Nigeria – The Baker of Maiduguri – 01/10/15

Since it was founded in 2002, the Islamist movement known as Boko Haram set itself against the Nigerian state. After its founder Muhammed Yusuf was executed in 2009 the rebel group was presumed defeated – but it roared back from the dead in 2010 after a mass jailbreak by its fighters. Since then it has pursued a full-scale insurgency in the country's northeast and beyond, spreading across the borders to Chad, Cameroon and Niger, attacking marketplaces and schools, often using suicide and truck bombs. It has abducted thousands of children, forced boys to serve as child fighters and killed aid workers. It has particularly targeted women and girls, as in the notorious mass kidnappings from a girls' college in Chibok, sparking the global #BringBackOurGirls campaign. The fear of its tactics – and those of the Nigerian army as it struggled to fight back – forced nearly two million Nigerians from their homes and nearly a quarter of a million out of the country. Katerina Vittozzi met one man trying to do what he could for the internally displaced.

'Have you heard about the baker and the IDPs [internally displaced people]?'

I thought it was the start of a bad joke when my Nigerian friend asked me that. We were having a drink in a hotel in Maiduguri, Boko Haram's former stronghold. But it turned out he was serious. He'd heard about a local bakery that had become a sort of refuge for people displaced by Boko Haram attacks, and thought I should go and find it.

Easier said than done. Maiduguri is a large city, the capital of Borno state, and I only had vague directions to the bakery. When I got to where I thought I should be, I didn't see any buildings that looked likely. No signs, no tell-tale sacks of flour, and the traffic fumes overpowered any possible scent of baked bread. But then I saw it: a building with rusting, corrugated metal walls, and on the roof a brick chimney pot emitting a thin line of smoke. Something was cooking.

Inside, Lanah Del Gashua, the baker, came over, holding his fist against his chest in a traditional, polite greeting. He was tall, in his late fifties and smartly dressed in a knee-length cotton shirt and loose trousers; it was Friday afternoon and he'd just come from the mosque.

Lanah's bakery was equally tidy – made up of one large main room with a cement floor and a brick oven in the corner. I was wrong – nothing was cooking there. The fire was dying down, the day's baking done. Piles of rectangular and round loaves of white bread were stacked on the wooden table, wrapped and ready for sale. On the far side of the room, I saw about a dozen boys lolling on raffia floor mats. I was told the youngest was eight, the eldest sixteen. Some of the smaller ones giggled and returned my two-handed wave.

Lanah and I sat on small wooden stools away from the oven (which was still giving out a fair bit of heat). Via a translator, Lanah explained that all these boys came here after Boko Haram attacked their homes. 'They are the sons of other bakers, from villages outside of Maiduguri,' Lanah said. Their fathers were among the thousands who had been killed or kidnapped by the insurgents.

He pulled out a small laminated certificate, printed in English. It stated that he was the Chairman of the Borno Bakers Association – an administrative post, but Lanah told me it made him feel responsible for the welfare of bakers and their extended families. They started arriving in 2012. Just a few people at first. But last year, Boko Haram intensified attacks in and around Maiduguri. More people started to arrive, looking for shelter, food, water and money.

Lanah said he'd helped find homes for almost 300 people, most of them women and children. Some were living in shelters in his garden; others, like these boys, stayed in the bakery. He gets no support from local government. What he was doing to help, he did alone.

More than two million Nigerians have been displaced during four years of fighting. Fewer than a tenth of them live in official government camps. Most rely on friends, family or virtual strangers like Lanah. But his resources are limited – his gently rusting surroundings attest to that. He can't offer these growing boys much more than a roof over their heads and endless bread to eat.

Lanah pointed to a small shack, about five metres square, in the corner, by the entrance. I thought it was a cupboard but it's where these fourteen boys sleep. A few clothes hung on nails in the walls. In one corner, someone had neatly rolled up the sleeping mats.

Usseini, fifteen years old, had lived there for two years. 'Do you get into fights over space?' I asked him, miming a few jabs. He laughed and nodded but said he always tried to get in early to secure his favourite sleeping spot, in the corner, by the room's only window. That was where he kept his little bundle of things, including an English vocabulary book. It was just a couple of tired-looking pieces of paper stapled together. He practised some words on me; numbers mostly, hellos and goodbyes.

I asked about school. Usseini shrugged and told me he hadn't been since arriving here. He wanted to go back though, once he got home. He wanted to study and become 'a big government man'. For now, those dreams were being kept alive in this shack, with this tiny bundle of English words and the generosity of a kind man.

Alex Duval Smith – Mali – The Northerners Next Door – 13/04/2013

There were many layers to the crisis in Mali. Complex relationships between farmers and herders, criminal ambitions, and the sense of a wider identity under attack all played a part. A constellation of extremist Islamist rebels, Tuareg secessionist groups and smugglers' syndicates were vying for control in the desert north of the country. After the Tuareg insurgents declared a breakaway state called Azawad in January 2012, and a military coup replaced Mali's central government in the southern capital, Bamako, France sent in its troops in 2013 – and helped the Malian army retake the whole country within a month. They were soon replaced by an African Union force, and a peace deal between the government in Bamako and the Tuareg rebels (though not the Islamists) was signed in 2015. The conflict is not over; rebel groups are still fighting and there have been several terror attacks on civilians and AU troops. Many people driven from their homes in northern cities like Timbuktu were quick to seek refuge in the south. But at home in Bamako, Alex Duval Smith noticed that accommodating the displaced was causing some domestic strain.

Before the military coup a year ago, my neighbourhood was popular with retired ministers, diplomats and expats. L'Hippodrome was known for its quiet, leafy streets and being handy for the French school. Many residents had quite a payroll of staff – guards, gardeners, cleaners and cooks.

My Malian neighbour Oumou, a single professional woman in her fifties, lived alone in a four-bedroomed villa inherited from her parents. Oumou has two brothers living in France. A feminist and staunch defender of Mali's status as a secular republic, she prefers to stay in Bamako, because there are still battles to be fought here – for freedom of speech, for women's and girls' rights. Oumou is like me. She smokes and drinks beer and wears trousers.

But a year ago came the advance of secessionists in the north of Mali, followed by their Islamist supporters. L'Hippodrome declined as foreigners left, rents plummeted and domestic staff were laid off. A local football pitch became a repair stop for long-distance lorries. The street next to mine turned into a sort of timber yard where broken furniture gets put back together. My local baker, having hit on hard times, has let his shop to a car mechanic. Now, collecting one's baguette involves negotiating a floor littered with oily black car parts and legs sticking out from beneath vehicles.

Oumou's life has changed beyond recognition. In October last year, her brothers in France sent cash to their aunts, who are farmers in Timbuktu, and ordered them and those they cared about to get on a bus and make the 500-mile journey to Bamako. Not just Bamako but L'Hippodrome and the large family house next to mine. Oumou was expected to accommodate eight people ranging in age from six to seventy.

You could argue that, living on her own, Oumou had three bedrooms to spare. 'It's not that I mind,' she says. 'But the people of the north are different. Their lives are governed by the seasons and the religious calendar.'

Coming from the desert – where they're used to communicating over vast distances – they don't talk, they shout. They don't laugh, they cackle uncontrollably. And they do all of this out of doors. I know. I live next door.

Oumou's pots and pans have been moved out of the kitchen into the garden, the natural place for farmers to cook over a coal burner. Her pet dog Coco is now obese, having given up her carefully-measured portions of pellets in favour of delicious scraps of food discarded around the garden. To call it a garden is an exaggeration; these days it's a yard of brown compacted dust where a big hole has been dug for household rubbish to be dumped in.

The relatives spend their day lounging on straw mats and cooking dishes that are delicious but take all day to prepare. And to Oumou's disgust, they've not sent their children to school. Yet their involuntary flight from the north could have offered the opportunity for the girl – lively, pretty thirteen-year-old Bentou – to learn to read and write.

This culture clash between educated middle-class Malians and their rural relatives is happening all over Bamako.

One of the peculiarities of Mali's war is that it has not spawned many camps for the displaced. There are one or two in the country and more over the border in Niger. But the vast majority of the estimated half a million people who have fled their northern homes have simply moved in with family members like Oumou. Malians say this is because of the country's tremendous tradition of hospitality. But sometimes when I see Oumou in our local bar, eking out a single glass of beer to avoid going home to the relatives, I think she must be wishing it were otherwise.

France began its military intervention in January and is already talking about scaling back troop numbers with a view to leaving. But few people believe the north will be safe to return to anytime soon – especially if the discredited Malian army is the one keeping the peace.

The relatives are bored and want to go back up north, where their priorities are determined by the seasons and the religious calendar. Oumou also knows that when they finally leave L'Hippodrome, one of their first priorities in Timbuktu will be to organise the delayed wedding of thirteen-year-old Bentou.

Tim Whewell – Central African Republic – Abandoned Mangoes – 05/04/14

Civil wars ravage the land and deform the human geography of settlements where different sides had once coexisted. With new fences, trenches, walls – or the burnt and abandoned remains of what were once homes – even a deserted town can speak eloquently of the people who once fought over it. In the Central African Republic in 2014, Tim Whewell saw signs everywhere that all was not well.

You can always tell where there used to be a village by the sudden profusion of mango trees in the middle of nowhere. They're my favourite tree: their leaves so much glossier, more deeply emerald, than those of any other. At this time of year, most of the fruit is still small, hard and green. It's difficult to imagine how, by June, the ground below will be covered in a rotting yellow squelch, the aroma more sickly than sweet. But it passes the time to try to imagine it, as you bump and jolt for hour after hour down the rutted ribbons of red dirt that pass for highways in the Central African Republic, and try to guess where those who once lived under the mangoes have moved on to.

There's always been plenty of movement in the wide belt of Africa where the savannah of the Sahel gives way to ever-denser forest. In Sango, a national language of the CAR, the word *kpetene* means, roughly, 'stay out of trouble'. I'm told that's what they sometimes called new settlements founded by families who suddenly decided to leave their home villages to escape disputes with fractious neighbours. But there are bigger migrations in train too. A century ago, when they ruled the region, the French encouraged some of the semi-nomadic Peul people – known in English-speaking countries as Fulani – to move south from Niger and Chad, to provide a better supply of livestock in Central Africa. The Peul are herders. They're also Muslim. So with the cattle came the Koran ... part of a slow southward spread of Islam that's continued ever since.

The farmer and the cowboy should be friends, but in practice they've clashed all over the world, all through history – from the Bible's Cain and Abel to Broadway's *Oklahoma!* And in the latest clash, in the Central African Republic, it's cowboys – the Peul – who've lost. Today, along those red roads, you don't need mango trees to tell you where people once lived. The empty shells of their houses and mosques are still standing, blackened and roofless. And you don't have to guess where the owners have gone. Many have been murdered, others forced to flee, by a savage militia claiming to represent the country's Christian majority. It's ethnic and religious cleansing on a massive scale – revenge on all Muslims, the militia says, for some atrocities committed last year during a Muslim-led rebellion.

This tragedy, barely noticed by the outside world, is about many things. The cauldron of hatred has been stirred by failed politicians who want to stage a comeback, and by the country's northern neighbour Chad, covetous of Central Africa's resources. But it's partly about jealousy between those who had political power but were poor – the Christian majority – and those excluded from politics who seemed slightly richer – the Muslims, Central Africa's main traders and herders.

For days and days on the road, I see no cows. Then, suddenly, scores of cattle emerge from the bush – massive, dewlapped, lyre-horned. But they're not being driven by the Peul who must once have owned them. Instead we're greeted by a group of scary young men waving machetes, bows and arrows, and home-made hunting rifles. Their chests are swathed in strings of little leather pockets containing magic bark and other charms they believe protect them from bullets. We've killed the Peul, they say – these are our cows now. They've also kidnapped several Peul women whom I glimpse huddled behind the herd. I also notice a tiny baby in a sling.

The young men are part of a militia you see everywhere along the roads – a force supposedly raised to fight highwaymen. Now they kill and burn under the slogan of 'Central Africa for the Central Africans'.

Suddenly, all Muslims are 'foreigners', even if they've been here for generations. Hundreds of thousands of them are now sheltering in refugee camps across the border, in Chad and Cameroon. But what about their cattle? Everyone worries there'll be a meat short-age. The farmers needed the cowboys; no one else can look after or slaughter livestock properly.

In Bozoum, a town which had thousands of Muslims until earlier this year, I'm told there are just two left. One's a madman. The other's a butcher, a tall, quiet man in a pink cap and gown who's been allowed – or maybe forced – to stay because his skills are needed. I sit with him in his courtyard under the mango tree – everyone has one – but he's too scared to tell me much about what's happened. He's very lonely, but he won't leave. 'This is where I belong,' he insists.

I look up at his mangoes. Maybe he will still be here to enjoy them in June. But the fruit of many, many other trees will go unpicked.

Thomas Fessy – DRC – In Lukweti's Quagmire – 26/11/09

Over the years the Congo's endless conflicts have drawn in numerous armies and guerrilla forces, some from far afield, some tiny local groups.

In the east of the country waves of battle swept through the same valleys and towns as different factions, and various government and UN forces, launched offensive after offensive. Each push usually displaced thousands of civilians and led to yet more killing, torture and rape. Many of the smaller militias had a reputation for bizarre battlefield superstitions, as well as wanton cruelty. The armed group known as the Mai Mai were in the thick of the worst fighting in 2009, when Thomas Fessy set off into the war zone.

The road to Lukweti is one of the worst I have ever seen. We carried our own planks to bridge the giant potholes but even they weren't up to the job. Night had fallen by the time we reached a zone where the rebels were known to operate. Then the car got stuck. There was nothing we could do to free it. After two hours, the three of us – a reporter travelling with us, our driver and I – had fallen asleep in the car when we heard the sound of people approaching.

We turned on our headlights and about ten armed men appeared out of the darkness, pointing their guns at us. But then we got talking. Far from threatening us, they were keen to help. They shook the vehicle this way and that for ten minutes and eventually got it out of the hole and back on the road.

We were on our way to meet a Mai Mai rebel leader who calls himself General Janvier. His men have been engaged in bitter fighting with government forces. But after our various breakdowns, it was now too late to reach his headquarters. We got as far as a tiny deserted village where the general has deployed about a dozen of his men. My colleague and I were given a narrow wooden bed frame to share. No mattress. As Janvier's local commander, Colonel Felix, put it: 'Now you are like us. We are soldiers – we sleep anywhere.'

The temperature drops fast at night in these high hills. Even though I wrapped my head in my scarf, I was still freezing. We woke up at first light. In the rising morning mist, we ate some more of the boiled yams we'd been given the night before. The sound of mortars, rocket-propelled grenades and machine guns echoed in the thick jungle around us.

We continued on our way amid heavy fighting. The muddy road

where we'd left the car had turned into a front line. Until then, the fighters we'd been with had seemed relaxed. They believed the government troops would not penetrate their lines. But suddenly that all changed and within a minute, it seemed, anyone who had a uniform put it on. They were ready to fight.

A traditional witch doctor approached the rebel soldiers and sprinkled them with water. This, the Mai Mai believe, protects their fighters and makes enemy bullets bounce off them. As fighting raged nearby, we ran into another rebel group consisting mainly of child soldiers. They'd attached leaves to their necks and heads in the belief this would make them invincible. But despite the good-luck charms, we encountered wounded fighters, barefoot, limping and leaning on canes, covered in blood. Government soldiers were attacking from three sides.

By now, we were with a mixed group of local Mai Mai and Rwandan-Hutu fighters who've become a well-established part of the fighting in the Democratic Republic of Congo. In another abandoned village, we met the man in charge of the rebels from Rwanda. He called himself Colonel Sadiki. He was calmly sitting on a bench receiving reports over a walkie-talkie, wearing a floppy dark-green hat and a red basketball jersey beneath. Every now and then he would sip a bit of palm wine from a gourd that hung from his neck. Colonel Sadiki had a child-like smile, but he clearly had authority over his troops. Despite the dangers all around, I felt strangely reassured, even though I knew little about this man. Was he responsible for any of the atrocities for which his movement has become notorious?

We kept walking ... sometimes dodging bullets and mortar shells. We hadn't eaten for nearly twenty-four hours when finally we stumbled upon a clump of sugar cane. We picked and chewed the long juicy sticks with relish.

Later in the evening, I couldn't figure out which was the strangest feeling: having my plate of beans served by child soldiers; squatting in a hut for the night knowing that its usual occupants were on the run, fleeing from the fighting; or falling asleep without knowing whether we'd be attacked. The morning after, as we were walking towards the no-man's-land between the two lots of fighters, we met a group of women who had been separated from their families. They were exhausted and fearful. A mother of four told us how she had been dragged into the forest by two government soldiers. They'd taken it in turns to rape her.

These are just some of the horrifying scenes of war in a place where the law is meaningless. I spent four days on the front line here and it felt like weeks. These women and their families have been trapped in this vicious cycle of horror for years.

Andrew Harding – DRC / Uganda – The Kony Scam – 17/03/12

Foreign interest in African affairs is notoriously fickle. While many nations' suffering goes almost unnoticed, there can be unpredictable surges of global outrage – all too often triggered by emotive, simplistic campaign videos. A powerful short documentary called Kony2012, produced by a supposedly charitable US group called Invisible Children, was one example: it was the first ever to rack up a million 'likes' on You-Tube, and more than half of all young Americans had heard of the video within three days of its release. Its subject, the brutal Ugandan 'prophet' and warlord Joseph Kony, was still at large – just as he was believed to be in 2020. When the video emerged, the former Congolese rebel leader Thomas Lubanga had just been found guilty of using child soldiers by the International Criminal Court in The Hague. Andrew Harding reported extensively on the activities of both men and explained the longer story which lay behind the intense, but short-lived, viral outcry.

I was hiding under a bed, in a priest's house, in the middle of a town called Bunia. It was 2003, and the Democratic Republic of Congo was living through a particularly vicious conflict. There were reports of cannibalism in the countryside. Rival militia gangs fighting for control of the nearby gold mines. Machete victims packed head-to-toe in the hospital. And at that very moment, the sound of heavy gunfire on three sides of the house.

The priest, a jovial Belgian called Father Jo, had seen it all before. As the bullets started to fizzle through the trees overhead, he lit another cigarette and went out into the garden. Just kids, he said, serenely. And he was right. Bunia was being overrun, yet again, by child soldiers.

For months we'd seen their blank, drugged eyes around our car at roadblocks. Weapons brandished like toys. Many of them reported, if that's the right word, to a warlord named Thomas Lubanga. Smooth and aloof, the besuited Mr Lubanga sought to portray himself as a politician. When I went to see him at his fortified compound, his chief of protocol spoke in hushed terms about 'the president's' schedule – then tried to stop us filming the bullet holes in the walls.

Nine years later and Mr Lubanga has just earned the uncomfortable honour of being the first person to be convicted by the International Criminal Court. He sat impassively as the verdict was read out in The Hague, still aloof, but now facing decades in jail.

Back in 2003, it was French paratroopers who ended the fighting in Bunia. Now in 2012 a foreign court is finally administering justice. Both events are a reminder that outside intervention can work in Africa, however imperfectly. But they also underline the failures of African leadership: how much better it would have been for everyone if the Congolese army had been able to stop the bloodshed and the Congolese justice system been able to put Mr Lubanga and many more like him on trial.

Perhaps that day will come. These are complicated times for Africa. The continent is changing fast, mostly for the better. Wars are ending, even in Congo. Economies are booming. Democracy – or something like it – is spreading, along with Chinese investment. There are still conflicts here, and failed states, and a persistent need for outside help. But many in Africa are seeking to renegotiate the terms of their relationship with the West.

If you are the parent of a teenager, there's a fair chance he or she came home from school in the past few days, breathless with outrage about a Ugandan killer named Joseph Kony. For the past two decades Mr Kony has run what amounts to a terrorist cult

– the Lord's Resistance Army. The LRA are responsible for the rape, torture, murder and abduction of thousands of children. A group of American activists made a film about Mr Kony, put it on the internet and more than a hundred million people have now clicked on it.

In 2004, I spent a lot of time in northern Uganda, tracking Kony's LRA and meeting their victims. I still remember the dogs skipping off into the bush with their spoils at the scene of a massacre. Two hundred villagers hacked to pieces. Their huts burned. A thirteen-year-old boy called Innocent had lost his entire family. We took him to a local orphanage in a town called Lira.

Today, Lira is peaceful. In fact, Joseph Kony's forces have not been active inside Uganda at all for six years. The armies of several neighbouring states are now cooperating to try to finish off what's left of the LRA – just a few hundred fighters. In Uganda, former child soldiers are being rehabilitated, communities reconciled. Progress.

The American film is not dishonest. But it shows none of this. It is outdated and simplistic, and many here find it deeply patronising. It perpetuates the idea that Africans are helpless victims, waiting for Western college students to ride to their rescue.

When the film was given a public screening in Lira this week, locals reacted with anger and disdain. I'm still in touch with the orphanage where we left Innocent. The man in charge emailed to say, 'We are at peace now. The film has brought back unwelcome memories. We are not happy with it.'

Help us by all means, he implied. But listen to us first.

Kim Chakanetsa – Zimbabwe – Making a New Plan – 31/05/18

Robert Mugabe hung on to the Presidency of Zimbabwe for a hair under thirty years, during which the country endured several brutal crackdowns on dissent and prolonged economic misery. But finally it happened: after a wave of protests, he was edged out of power in

November 2017, though he got his revenge in first by officially resigning. When Emmerson Mnangagwa became president of Zimbabwe, his supporters brandished cuddly crocodile soft toys in celebration – a reference to the nickname he'd earned for his political cunning: the Crocodile. Mnangagwa promised more jobs, less corruption and restored international ties – including taking Zimbabwe back into the Commonwealth. But returning to the country, Kim Chakanetsa wondered how far the promised 'new dispensation' was likely to go.

One of the first things I notice as I wheel my bag into the arrivals hall at Robert Mugabe International Airport is that the man himself is nowhere to be seen. Gone are the framed portraits that used to decorate the country's schools, hospitals and hotels. As I exit the sliding doors, I notice that his portrait has been replaced with that of the country's new leader, Emmerson Mnangagwa. The portrait is almost identical to Mugabe's – sharp suit, bold tie, raised chin and an enigmatic half-smile.

When the coup that was not to be called a coup reached its final act last November, it was the portraits people went for first. In one of the many celebratory videos posted online, four women wrestle a picture of Mugabe off the walls of a swanky hotel in Harare. A brief tussle with staff ensues: the women become more animated, cheering each other along, until one of them snatches the portrait away, sprints out of the hotel and hurls it to the pavement and begins to jump on it together with her friends as they chant *'Ayenda! Ayenda!'* He's gone! He's gone!

As I make my way into the city, I ask my driver if anything has changed since. He laughs: 'No, nothing!' He tells me that, for him, every month remains a struggle to see if he can make the nursery payments for his daughter. He relies on two jobs; one hasn't paid him in five months but he shows up anyway, hopeful that this will be the month that he finally gets his back salary. 'You know, you just have to make a plan,' he smiles. You'll hear that phrase a lot in Zimbabwe – making a plan. It reflects a hustling spirit that accepts things as they are and tries to find a way around them.

As we weave into the city centre, dodging occasional potholes and packed commuter buses, I realise that we've been in the car for more than twenty minutes without passing a single police road-block. Now that *is* a change. Most Zimbabweans have stories to tell of overzealous police officers racking their brains to find any pretext for an on-the-spot fine. I've heard of fines being issued for reflective stickers that weren't shiny enough, for dirty vehicles and even for torn car seats.

Cracking down on the roadblocks has won the new government some measure of goodwill. And they are going to need as much of it as they can get. The economy is still in terrible shape. Unemployment is high and there is a biting cash shortage. One afternoon, as I waited to pay for a few groceries in a local supermarket, I took out a US $10 note. A woman who was also waiting in line rushed up to me. 'Please,' she asked, 'can I use my card to pay and you give me the cash instead? Please?'

The cash crisis is painfully visible in the numerous queues that stretch and curl around the city. Outside one bank on a Saturday morning, a woman holding on to a ticket numbered 146 explained to me that she had been there since 5 a.m. 'I came yesterday, but I got nothing. It's really a problem,' she sighed. 'It's your money and you've worked hard for it and you can't even get it out. Something needs to change.' Another woman thought things were already looking up. 'Before, you could only get about thirty dollars out of the machine,' she explained, 'but now you can get fifty dollars. I think Mnangagwa is trying, we just need to give him a chance.'

For the most part the president seems to be making the right noises. Earlier this year, he made a foray into social media with a slightly awkward Facebook Live video from his home. Apart from some snide remarks about his sofa covers in the comments, this 'man of the people' move was applauded. It's not something you would imagine a certain former long-time leader doing.

But there are others who see only a Mugabe 2.0. A man who lived permanently in the shadow of Mugabe and is somehow an extension of his rule.

The truest test will come in the elections which are due to take place in July. In the past, Zimbabwean elections have rarely been free or fair. ('The winner is going to be the winner,' is how one cabinet minister confidently put it.) But Zimbabweans have been burnt before and it will take a lot to convince them. The government will have to do what citizens have been doing for years now. They'll have to make a plan.

Hugh Sykes – South Africa – World Cup Disappointments – 28/05/10

There's a view that after its first multiracial elections in 1994, South Africa coasted for years on its apparently miraculous escape from political violence. Its new ANC government, assured of international goodwill and bolstered by one of the world's most progressive constitutions, could hardly put a foot wrong. Sure, there was the challenge of the HIV pandemic, and the odd case of corruption to deal with, but this was the Rainbow Nation – capable of defying strategic gravity, a place where old sins could be redeemed. Hosting the FIFA World Cup in 2010 was a perfect showcase for this version of events. But Hugh Sykes discussed real tensions still roiling the country not far below the picture-perfect surface.

In a community of shacks on a hillside near Johannesburg, a man complained to me: 'We didn't like apartheid, but some things were better under apartheid than they are now.' In a community of shacks on a hillside near Durban, a man complained to me: 'Life here under apartheid was bad, but now it's more bad.'

I felt unsettled hearing this. It seemed like questioning a sacred belief – that apartheid was an unmitigatedly, 100 per cent evil system. But there's less idolatry here now, as it dawns on most people that the new South Africa is still scarred by extreme poverty and high unemployment.

Of course, Nelson Mandela continues to be lauded as the hero of the liberation of black South Africans. But he's also being criticised for changing the direction of the South African economy from active state intervention to neo-liberal free-market economics.

During his presidency, the government switched from RDP to GEAR. RDP was the statist, interventionist Reconstruction and Development Programme. GEAR is more market-orientated. It stands for Growth, Employment And Redistribution. RDP promised paradise: clean water, mains drainage, land redistribution and a million homes – all within five years. But paradise didn't come: the economy of South Africa simply couldn't bear the cost. So the finance system switched to GEAR.

Part of the thinking was that it would help to develop a substantial black middle class, whose taxes would then trickle down to the poor. The middle class did develop, but the problem with trickle-down is that it's just that – a trickle. Millions of South Africans still live in shacks and have to go out to public stand-pipes to fill up containers with drinking water. Their homes have no proper lighting or security from burglars, and the rain and dust get in.

A former African National Congress activist, Bricks Mokolo, told me it's still very hard to criticise the government here. He says everybody has been, as he put it, 'made to love the ANC, made to love Nelson Mandela' and 'made to feel small' if they dare to complain. Mr Mokolo tells me angrily: 'I didn't just wait around for Nelson Mandela! I too fought for my freedom. I was tortured in an apartheid jail.' He was tortured so brutally that prison officers thought he was dead – after leaving him in a mortuary fridge overnight, they dumped what they thought was his dead body in a field.

Mr Mokolo says that housing was better under apartheid than it is now. He calls the new houses which are being built all over the country an insult, because they're significantly smaller than the old matchbox homes that the apartheid government built in the townships. 'The ANC government now,' he insists, 'is simply an extension of the apartheid government. There are still townships twenty years after liberation.' His conclusion: 'There were places for blacks in those days. Now they are the same places. They've just changed the word. They've changed "black" to "poor".'

I have heard this radical outlook expressed much more often here than on previous visits. A wave of strikes over the past few weeks is evidence of mounting frustration – despite the undeniable success of the growing black middle class in their smart homes and fast cars. 'But what about us?' say the rubbish collectors who were on strike for ten days in April, or the train and dock workers who were out for more than two weeks this month.

Mineworkers were about to go on strike too a few days ago and were only stopped by a court order. If they had stopped work, there could have been a threat to power supplies ... during the World Cup. Stadiums are now taking precautions, installing generators, and back-up generators. This reminds me of the last World Cup, when I was in Baghdad: there, power cuts blanked out TV sets during some of the games. That surely can't happen here ... can it?

Pumza Fihlani – South Africa – Cape Town's Water Crisis – 01/02/18

'The fairest Cape in all the world': the extraordinary beauty of Cape Town and its setting still draws visitors (and would-be settlers) from around the world to post-apartheid South Africa. But despite the crowds of tourists and the official desegregation of all public space, it's still riven by inequality and the pressures of gentrification, too. In 2018 it also became a stark test case for a global problem, as it teetered on the verge of becoming the world's first major city to run out of water. After three years of drought, its reservoirs were drying up and local government warned that taps might have to be turned off. Across Africa, the UN noted that up to 300 million people didn't have access to clean drinking water. In the countdown to 'Day Zero' – as the looming dry-tap moment was called – Pumza Fihlani pointed out that some were likely to cope better than others.

It's four in the afternoon. Scores of men and women are standing in a queue to collect water, each carrying a plastic bottle or two – some, a large bucket. They're only allowed to collect twenty-five litres of water here at a time. A security guard is on hand to make sure that everyone behaves. The crowds sometimes get a bit restless when the queue seems to stall – but for the most part everyone waits their turn.

In the shadow of Table Mountain, residents are worried this could be their new normal. People are being urged to use fifty litres of water per person per day, down from eighty-seven litres previously. They've been asked to limit their showering time to just two minutes, to flush the toilet only twice a day – and to re-use grey water around the house. But water use is hard to police. It turns out that, so far, only a third of Capetonians are following the new rules.

There is no doubt that the situation is dire, but there has been little sympathy from people outside South Africa's prestigious wine-producing province.

Cape Town is a strikingly unequal place. There is luxurious Cape Town, with its rolling winelands, five-star boutique hotels and award-winning restaurants, manicured golf courses, and homes whose grounds could accommodate a football field – the sort that have Olympic-sized swimming pools. Then there is grim, congested Cape Town – whose shacks are home to millions, with narrow streets where sewage flows freely, and only a handful of rarely serviced outside toilets for an entire community. Home to townships where a shower of any length at all would be considered a luxury. You don't see *that* Cape Town on the postcards. The two worlds barely intersect – the race and class divides enforced by apartheid are still evident today.

The city is run by the opposition Democratic Alliance, a predominantly white party, so it's been perceived as being geared to promote white interests – the haves. The authorities would deny that, but there is a reason this place has been nicknamed the Republic of Cape Town – a country within a country, and perhaps one out of touch with the issues affecting the rest of South Africa. Amid the water shortage, local government has claimed the worst culprits are big business and the wealthy – they've simply refused to scale down their consumption.

So why the nonchalance from fellow South Africans? Well, millions of people here have either grown up without clean running water, or had to ration what little there was to survive day-to-day. In many rural areas, standing in line for water is the norm, the usual routine, unless there's a river nearby. If there is a source of piped water, it's often a communal tap shared by hundreds. This is also how people living in the Cape's shanty towns and townships have lived for decades.

So why has the Cape Town drought story so dominated national and international headlines when similar crises in remote areas have barely made a splash? Maybe it's because the water shortage, or having to cut back on water usage, is now affecting those accustomed to privilege. Poor people across the country have been living in 'day zero' all their lives, but their cries barely make the news.

I remember growing up in the Eastern Cape, a beautiful but underdeveloped province. There would be days when we would have to walk for hours, pushing wheelbarrows piled with twenty-five-litre plastic containers to collect water from a stream, when the municipal supply failed. Each person of the five in our household would be allocated twenty-five litres of water a day for everything.

We lived in a small town, in a nice enough neighbourhood – and usually had clean running water in the house, two flushing toilets, lights and tarred roads – but even as a young girl I was aware that this wasn't the norm for people everywhere. The trips to the stream were nerve-racking for me: the communities who relied on that river didn't take kindly to having 'suburban outsiders' stealing their water and they'd let us know it. At times my younger sisters and I would wake up really early, when we thought the locals were still sleeping, to avoid any confrontation. They resented our ability to tap in and out of poverty, the luxury of being able to return to our real lives.

And so the idea of people standing in line for water does not seem as shocking to me as it is for many Capetonians. This is how generations lived before, including my mother's. A life many South Africans are still trapped in. The difference is they have already developed their own ways of coping.

In Cape Town, that deeply divided city, maybe water could be the ultimate equaliser after all.

Tristan McConnell – South Sudan – The Threat Is from Within – 15/10/14

The birth of a whole new nation, a sovereign state officially recognised by the UN and formalised by treaty – surely one of the purest possible examples of real news. South Sudan became the world's newest country in 2011 after a long and exhausting struggle to break away from Sudan, as defined by (and ruled from) Khartoum. When the people of the South voted in a referendum in January of that year, the result was stark: 98.83 per cent wanted full independence. It appeared a dream come true: the final reward for countless lives and lifetimes of sacrifice.

Barely three years later, the initial euphoria was in tatters. A country 'smashed by war', which already had some of the worst levels of poverty, malnutrition and illiteracy in the world, had ruined itself even further after different factions within the government fell out, and their militias divided along ethnic lines. Tristan McConnell saw how comprehensive the destruction had been.

When I last visited Malakal it was a time of hope and trepidation. In late 2010 South Sudan was on the verge of realising its dream of independence after decades of civil war. There was uncertainty, of course, but at the ramshackle port on the Nile there was joy too. Barges that had steamed upriver for days arrived stuffed with people lugging their possessions returning to vote in the coming independence referendum. They disembarked, greeting long-lost relatives with smiles, laughs and a traditional clasping of the shoulder. They were home and, after generations of colonial rule and Arab subjugation, that home was soon to be theirs.

Malakal was a cultural crossroads and trading centre close to the north–south border. There was a mosque and a church. In the busy market, northern Arabs and black southerners shopped and sold goods side by side. It was also prosperous, by the region's low standards: there were imported items, overpriced hotels, lots of aid workers and a large teaching hospital.

All that is gone now: the hope, the joy and the activity along with much of the town. In the new civil war that has torn at South Sudan since December, Malakal has changed hands six times. With each shift of control it has been bombed, shot up, burned, emptied, looted and abandoned. Today it is a ghost town inhabited mostly by soldiers. Long grass is already reclaiming the outskirts, flowering creepers hang like bunting from disused power lines, the walls of abandoned homes and buildings are covered in beetling cracks and splitting apart.

To meet ordinary people – which is to say, ones without guns – you go to the nearby United Nations base where 17,000 of them have sought shelter. When I visited in September there was fighting to the north and to the south of Malakal. When the rainy season ends, everyone expects another assault on the town, and another change of hands.

As I drove through town I saw a shop I remembered. Back in 2010 a young Darfuri trader called Yusuf had proudly told me how he had built his successful second-hand clothes business here from nothing. Now it was nothing again: the shop was empty, the doors bent off their hinges, the walls charred. I hoped Yusuf had escaped.

Even the hospital had been looted and burned. A doctor told me they had found eleven patients shot dead in their beds, including a mother and her infant child.

Like South Sudan, Malakal has suffered many turbulent periods during decades of north–south civil war and years of tentative peace, but David Koud told me the recent fighting was the worst he'd ever seen. I met David in a place called Wau Shilluk. It used to be a sleepy village downriver from Malakal, but the population increased more than sevenfold as people fled the fighting in a flotilla of vessels. One boat capsized as it crossed the Nile in January drowning 300 people at a stroke.

David is a sombre, softly-spoken man in his mid-thirties. Before the war he had been a civil servant in Malakal with a television set, two cars and two small houses. Life was good, he told me. When the war reached Malakal he lost track of his wife and two young sons. A week after fleeing town, David returned to find the body of his brother lying dead in his bed. Also in the room were four other men, all of them shot dead. David didn't know who they were.

It was months before he found his family again. They had escaped and were living with relatives in a village with no phone reception. In between, David said, he had lost his mind, become a madman, sure in the knowledge that everyone he loved had been taken from him.

As we spoke in David's makeshift home of sticks and plastic sheeting, his six-year-old son practising handwriting and his four-year-old using his dad as a climbing frame, David told me he wanted nothing from South Sudan any more, except to leave. This latest civil war – so close, so personal, so brutal and so destructive – has vanquished hopes that survived decades of previous fighting.

Back in 2010 I had met a courtier to the Shilluk tribal king in his quiet, neatly swept compound in Malakal. We discussed what the future, and independence, might bring. Samuel Aban Deng looked forward to the freedom but warned that it would not be a panacea for his country's troubles. 'The bigger threat to an independent South Sudan is not from the outside,' he told me, 'it is from within.' How right he was.

Alastair Leithead – Ethiopia Rising – 26/08/17

Even as some African states squandered their human resources and dashed their citizens' hopes, others were on a different trajectory altogether – riding the growth of a middle class, and some of the world's fastest-reducing birth rates and fastest-rising standards of living. Far from the apocalyptic visions summoned by the alarmists, there were analysts who witnessed lands of demographic and economic opportunity. Alastair Leithead saw just how different today's Ethiopia is from the one he'd grown up seeing on television, and heard what its leaders had planned for a brighter future yet.

Driving up over a hill out of Addis Ababa, on a modern but eerily quiet six-lane highway, the hollowed-out concrete shells of seven-storey apartment buildings appear out of the green fields, filling the horizon. The sun lights up their stark silhouettes, cutting through the coarse square holes where individual flats should be and illuminating the narrow streets between buildings. Tapering off as far as your eye takes you, they're like the vast ruins of a post-apocalyptic city from which all the people have fled. A deserted railway runs alongside the road – its overhead power lines dormant. But these aren't relics of some past tragedy: they're monuments to Ethiopia's future.

The people just haven't arrived yet. The apartments are government-built social housing under construction – a grand-scale concrete production line. The road is just one small stretch of a vast new network spidering across the nation. And the railway is waiting to whisk imports and exports between the capital and the coast: a gateway for the ambitions of a country which now oozes strength and self-confidence. Betting, as so many now-successful countries did before it, on massive building projects, on an agricultural revolution, and on industrialisation to drive an economic boom and catapult the country out of the nightmare that was 1984.

It was more than thirty years ago, but talk of Ethiopia abroad still conjures up images of famine: starvation; malnourished, dying children; vast tented desert camps; pop stars begging us to Feed the World. That Ethiopia is gone, but is far from forgotten – certainly not by a government determined to be a powerhouse in Africa, an example to the continent of how to take an exploding population and put it to work. Last year the country suffered a drought far worse than that in 1984, but the outcome was very different. The government spent more than 700 million of its own dollars, buying and transporting food, keeping its people alive.

Every morning at the international airport in Addis Ababa, a gleaming line of tailfins, each bearing the green, yellow and red of the Ethiopian flag, disappear off into a heat haze. Billions of dollars' worth of aircraft ready themselves to head east to Asia, north to Europe and to every corner of the continent. It's like an African Dubai, where the Emirates airline put a city state firmly on the international map. Ethiopian Airlines, too, is a symbol of a country on the rise.

So is one of Africa's largest industrial estates. Recently, under a big tent, the prime minister officially opened Hawassa Industrial Park, south of the capital. It's one of more than a dozen such projects being built across the country, this one focused on textiles. In a square surrounded by a gridiron of factory units, a state-of-the-art water purification plant and even its own textile mill, sat executives from some of the world's biggest clothing companies. The theory was, build it and they will come. And come they did – from America and China, India and Sri Lanka – because the government built the facilities and offered an alternative to Asia.

The executives all said similar things: 'We had a good, close look across Africa and realised the best place to be was Ethiopia.' Labour is cheap and plentiful, a growing population is eager to move from agriculture to industry. 'The jobs don't pay that well,' the workers I found cutting, sewing, ironing and packing on the production lines told me. 'But it's not just about money,' another said. 'Opportunities are being created for people like me across the country.' The factories are seen as a pathway out of poverty, providing careers, education, and the skills needed to start their own businesses in future.

But there is a cost beyond just debt. On the road from Addis Ababa, a few burned-out trucks litter a small stretch of highway. They're a reminder of the protests that erupted last year against a government less concerned about human rights and freedom of speech than about remaining Africa's fastest-growing economy. The country is still under a state of emergency, and hundreds of political opponents have been jailed.

Sitting at his desk in a sharp suit, Arkebe Oqubay, the architect of the industrialisation policy, is steely and determined: utterly confident in his approach; unapologetic about the country's priorities. Yes, they could be more inclusive, the prime minister's special adviser concedes, but without a richer country, what's the point? In cities, he says, the population is growing by about 5 per cent annually, so they need to create nearly one million jobs every year just to keep up. As in China, freedoms may come later.

The memories of 1984 are being buried – deep below tens of

thousands of tonnes of poured concrete, great sweeps of tracks and tarmac, and beneath a state of emergency which gives the government the power it says it needs to force through its ambitions.

Mary Harper – Somalia – My Friend the Dry-Cleaner – 18/08/2018

Over the decades of overlapping civil wars since 1991, a Somali diaspora was sown and grew across the world, from Minnesota to Sweden to the UK. More than a million exiles, refugees and émigrés, many with skills, connections and ability sorely needed in Somalia – and many of them yearning just as sorely to come home and help rebuild it. People like the entrepreneur Mohamed Mahamoud Sheikh Ali, who had often helped the BBC's Mary Harper on her regular visits to Mogadishu . . .

The last time I saw Mohamed, he gave me flowers. He chose one of the biggest and most colourful bunches from the display in his florist's shop. Next door, machines whirred away at a dry-cleaner's – which also belonged to him.

Stopping off to see Mohamed was always one of my favourite things to do when I visited Mogadishu. While the world's media spoke of famine, pirates and suicide bombs, he quietly and determinedly got on with his life, bringing what many would see as entirely normal, mundane services back to his country.

He also encouraged others. He set up a community of people involved in start-ups, and became something of an inspirational figure – but always remained modest. Lots of brave new businesses have sprung up, from the young man with a motorbike who has started a food delivery business to the girl who has set herself up as a mechanic.

Visiting Mohamed was not entirely straightforward. As for all my other appointments in Mogadishu, I never fixed a precise time. Sometimes I would just show up outside the metal gates of the Somalia Premium Laundry on the busy Maka al-Mukarama road. I always travelled there with heavy security – at least six bodyguards in one vehicle, a couple more in the other. It is best to be unpredict-able; people say phone calls are eavesdropped on and that there are informants everywhere.

But taking care is no guarantee of safety. Earlier this month, Mohamed was driving in his car in full daylight in a heavily guard-ed area known as Kilometre Five. Two men appeared and shot him. This unassuming but influential young man died later in hospital. So far, his murderers have not been caught.

Now Mohamed won't be able to realise the other dreams he told me about. Of opening a gym, a playground for children, of growing all the flowers for his shop in the fertile fields of Afgoye, not far from Mogadishu. He dreamt of greening the city, and had already brought in flowering trees to plant there.

Somali social media was soon awash with comments from people whose lives he had touched. Many were accompanied by the hash-tag #WeAreNotSafe. Three days after he died, a rare demonstration was held in Mogadishu. Young people wearing white headbands held up banners emblazoned with phrases like 'Stop Killing Youth'. They asked how and why people like Mohamed were being killed, and why nobody was being held accountable. On that very day, a car bomb exploded outside a restaurant in the city, killing at least three people.

I first heard about Mohamed when a friend called to tell me about someone who'd given up a safe, well-paid job in Dubai to return home and set up the first dry-cleaner's in Mogadishu for more than two decades. I thought this would make a great story: Somalis like Mohamed, who'd lived in peaceful countries abroad, coming home to rebuild their nation.

The dry-cleaning element also appealed. Whenever I flew out of Mogadishu, I'd see politicians and businessmen board the plane carrying vast piles of dirty suits to be cleaned in neighbouring coun-tries. Once, at a summit held in Ethiopia's most luxurious hotel, I was astonished to see Somalia's top politicians bustling down the corridors with armloads of freshly dry-cleaned suits. Mohamed had spotted an excellent business opportunity.

His murder has thrown up questions about the nature of violence in Somalia. About who is killing who. People often rush to blame al-Shabaab, which for more than a decade has spread terror in Somalia and beyond. But the jihadists are not the only killers. It could be a politician who doesn't like what you do or say, a business rival ... or it could be because of a property dispute, or plain jealousy. People are quick to reach for their guns. I have been stuck in Mogadishu traffic jams where the guard in my car has rolled down the window and fired live bullets into the air, just to get the other vehicles moving.

Mohamed was not the only rising young star to have his life cut short. Abbas Abdullahi grew up in a refugee camp and was named a government minister last year. He was shot dead accidentally – by the attorney general's bodyguards. Young journalists are regularly murdered. It seems strange that, despite all the billions spent on security in Somalia, the presence of tens of thousands of African Union troops, US drones and special forces, there is still little protection for people like Mohamed.

I keep catching myself thinking about him – about our friendship, his welcoming gap-toothed smile and his unswerving commitment to making life better. I wonder about his businesses. Are they standing empty now, the dry-cleaning machines quiet and still, the flowers wilting? Is anybody watering the pots of plants which he tended so carefully, and sold to people trying to bring a bit of brightness into the homes and businesses they are rebuilding in Mogadishu?

Part III:
THE MIDDLE EAST

Arab Spring

Paul Adams – Egypt After Mubarak – 19/02/11

The speed with which the mass protests broke out across North Africa at the start of 2011 took institutions, as well as individuals, by surprise. After demonstrations rocked Tunisia, the intensity of protest in Egypt managed to dislodge Hosni Mubarak – who'd ruled ever since the assassination of Anwar Sadat in 1981 – in a matter of weeks. The violence of the security state, as well as stifling censorship and economic stagnation during his regime, had embittered so many for so long that they lost their fear of speaking out. The Supreme Council of the Armed Forces took over, promising multi-party elections within six months, while Mubarak was placed under house arrest, to face trial before the year was out. Paul Adams revelled in the atmosphere, but also felt Egyptians' forebodings.

Just over a week ago, I watched as the waterfront in Alexandria erupted in wild euphoria. Not on the scale of Cairo's Tahrir Square, but just as exuberant. Fireworks. Shots in the air. A man leaning out of a car window, setting fire to the spray from one aerosol canister after another. It went on and on for mile after mile.

The following morning, under leaden skies and with a chilly wind blowing off an uncharacteristically grey Mediterranean, I walked along the beach to canvass opinion. And gradually – inevitably - it all started to appear a lot more complex. As of course it should.

Walid the fisherman couldn't have been happier. Thirty years of what he called 'a semi-occupation' had finally come to an end. Further along were a husband and wife sitting in deckchairs, like a doughty British couple taking the air at Margate in April. Mohammed said the old president had brought it all upon himself, appointing cronies who wrecked the economy. But his wife, Sabah, said Mubarak was irreplaceable. I wandered on to a café, where two young men made no bones about their concerns. 'The Muslim Brotherhood is trying to take over,' they told me, 'and if that happens, Egypt is finished.'

And so it went on, as I made my way back up the Nile Delta towards Cairo: ambiguities and contradictions piling up all the way, like belongings piled precariously on the back of a battered pickup.

At Kafr el-Meselha, Sami Abdul Hafez chugged towards me on his motorbike as we tried to locate the house where Hosni Mubarak was born. Sami – who called himself 'the poet', without ever quite explaining why, and who batted away swarming children with alarming vigour – explained that Mr Mubarak had turned his back on his home town and thus deserved little sympathy.

A teacher called Admad spoke, as so many do here, of having his human dignity restored. But others feel their country's dignity has somehow been sullied. Ashraf, a softly spoken English teacher, said this was no way to treat the father of the nation, even if he had made mistakes. What kind of image was this to present to the outside world?

And a group of angry men demanded that we hand over our film and threatened to destroy the car. Two of them escorted us out of town. Before leaving, they leaned in at the windows to explain that what had happened to the president was disrespectful – that the youth who led the revolt simply had no manners. They clearly felt that we were little better and suspected we had come to gloat. Once they sensed we understood, they smiled, almost apologetically, and said we were welcome in their town.

Back in Cairo, the blackened hulk of the ruling party headquarters looming over the Nile, and the tanks still parked nearby, tell you that something powerful has happened. The young activists, busily adding a lick of green paint to the ironwork on the bridges over the river or whitewashing the kerb, are showing in whatever way they can that they want a better, less shabby future.

But little is clear in this post-Mubarak era. The revolution – if that's what it turns out to be – has shaken the system but not brought it crashing down, let alone replaced it.

In countless state-run institutions up and down the land, textile and steel factories, transport and media companies, and even the stock exchange, managers and employees are locked in a tussle over the spoils. What does it all mean and how will it change their lives? Will it make them freer, or put a little more bread on their tables? Will the army really act as midwife to the new Egypt, or merely cement its own authority? It's tempting to look at the decay and chaos of this great, overcrowded city, its sclerotic bureaucracy and rampant corruption (to say nothing of the lack of convincing leaders waiting in the wings) and conclude that Egypt doesn't have a chance.

The winter revolution is a work in progress. There's a constitution to rewrite, political parties and elections to organise, and mindsets to overcome. But for all his faults, Hosni Mubarak left behind him a large, sophisticated middle class yearning to do something with their education. The revolution has given them an opening, and if rather a lot of people are now rushing to take advantage, then perhaps it's no worse than four lanes of Cairo's honking, madcap traffic, all aiming for the same exit.

Jeremy Bowen – Libya – The Colonel Is Confident – 03/03/11

Everything about Colonel Muammar Gaddafi seemed larger than life: his ambition to be a leader of the entire African and Arab world, his defiance of the West, his grand engineering projects. But his image had been magnified by a cult of personality which had shored up his control of Libya since the 1970s. He was the 'Brotherly Leader' and had no time for anyone else hoping to lead; when youth-led protests like those in Tunisia and Egypt erupted in Libya in February 2011, his contempt was absolute. In a televised speech he urged his supporters to root out and attack any such 'cockroaches', promising to 'cleanse Libya house by house'. Security forces used live fire against demonstrations in the cities of Tripoli and Benghazi, killing hundreds of people. Despite years of slowly improving ties, American, British and European leaders

denounced him, and the stage was set for another dramatic confronta-
tion between Gaddafi and the West. Yet Jeremy Bowen found the man
at the centre of it all surprisingly relaxed.

Even on the best hinges that Libyan oil money can buy and German
automotive engineering can create, the doors of the armoured
BMWs they had sent to take us to see Colonel Gaddafi were still
very heavy. They shut with a satisfying clunk.

I thought of the bone-shaking, boiling hot or freezing cold
bullet-proof Land Rovers we used in the war in Bosnia nearly
twenty years ago, and settled back into a seat covered in thick,
supple tan leather. I reached up to touch the roof: it was lined with
chocolate-brown suede.

'It's a James Bond car,' the man in the front passenger seat said
proudly. I don't know if it had an ejector seat, but there *was* an AK
47 assault rifle in the boot. The BMWs pulled away – soundless,
smooth, and going, I assumed, to some kind of tent in the desert.

The streets of Tripoli, a city which even the government
spokesman here describes as boring, were surprisingly busy. That
is significant. Over the years I have been in many unstable, violent
places. When regime change is in the air the streets are always
empty. During the last few weeks of President Mubarak's reign
in Egypt, even Cairo's notorious traffic problem was solved as so
few people wanted to risk going outside. I wouldn't describe late
afternoon in central Tripoli as rush hour, but plenty of people were
out and about.

The cars slowed down outside the main Libyan broadcast centre.
Was the meeting going to be in a TV studio? The man in the front
seat, who had arranged the interview, turned back. No, he said over
his shoulder, it's a restaurant.

The floors inside were made of glass, covering a shingle beach
of hard white pebbles. Colonel Gaddafi arrived punctually, dressed
in a russet-brown robe and wearing a pair of Ray-Ban aviator sunglasses. He was alert, cheerful and focused. In the interview he was
defiant.

The UN resolution against Libya was illegitimate, he said. Why didn't they send a fact-finding mission before they passed it? Al-Qaeda was responsible for the trouble in the streets. The West wanted to recolonise Libya. He had no assets abroad to freeze. 'I have only a tent,' he said.

A couple of years ago I was in New York City when Colonel Gaddafi made a speech at the United Nations. He delivered it at more or less the high point of his rapprochement with the Western world. He had even been at the G-8 summit in Italy that summer. A lot of the press coverage of his visit focused on his staff's efforts to find a place to pitch his tent. In the 1980s President Ronald Reagan called Colonel Gaddafi 'the mad dog of the Middle East'. In New York he was having disputes with suburbanites who didn't want him in their back yard.

The kindest thing you could say about that UN speech was that it was rambling. But he wasn't like that in the trendy-looking restaurant overlooking Tripoli's port. He answered the questions that were asked of him, and I'm sure that to the colonel and his supporters his replies seemed entirely logical.

To try to understand anyone, you have to think about how the world looks from their angle. Colonel Gaddafi's is unique. He has been in power since he led a coup in 1969, when he was in his late twenties. Photos of his early meetings with allies show President Tito of Yugoslavia, Gamal Abdul Nasser of Egypt – major world figures of the Fifties and Sixties. The colonel has been in power for almost all his adult life.

All political leaders live in a bubble. It's safe to say that Colonel Gaddafi's is more hermetically sealed than most. I suspect that after all these years there are plenty of yes-men around him. He told me that his people loved him, and would die to protect him. It sounded like one of the assumptions on which his unusual life rests.

He admitted that people had been killed in the disturbances – but only because they attacked the security forces, who acted in self-defence. The Libyans do admit they have bombed ammunition dumps, but deny they have used the air force against their own people.

For a man under pressure, Colonel Gaddafi seemed very relaxed.

Perhaps he feels more comfortable back in his old role as the West's bogeyman in the Middle East than he did embracing Tony Blair and Silvio Berlusconi and the other leaders who, for a brief moment, saw Libya as a new, benign, even pro-Western frontier in the Arab world.

When the colonel left, I thought I saw a spring in his step.

Andrew Harding – Libya – Benghazi: We're All Volunteers Now – 14/05/11

Benghazi, in the east of Libya, had a historic rivalry with the capital Tripoli throughout most of Colonel Gaddafi's long rule. As the seat of the former royal family, the Senussi dynasty, it chafed at being labelled the country's second city and was an epicentre of the protests which broke out in 2011. As Gaddafi clung to power, the armed rebels trying to overthrow him gravitated there – not least to shelter within the zone of air cover provided by NATO. The European Union announced it would be opening an office there, while David Cameron invited the rebels to open a formal office in London. Andrew Harding sensed a feeling among people in Benghazi that their time had arrived at last.

I'm strolling along the blustery seafront in Benghazi. A middle-aged man rushes up to me. He grins. Then starts to pant heavily. It takes a few awkward moments before I work out that he's miming: he wants to tell me that Libya was suffocating, but now people here can breathe again.

Colonel Gaddafi's leering caricature is scrawled on the walls around the harbour. A captured tank points its barrel out to sea. On a tent beside it, someone has painted the words NO FEAR.

There is a giddy sense of pride and optimism here, in the city where Libya's revolution began. 'Look at the pavements!' says Molly Tahouni proudly. She's a young academic who packed a rucksack and rushed back from a life in exile in London to help with an uprising that she had confidently predicted was still years away. 'How wrong I was!' she says, and points down to the kerb. Some young children roped off the road here yesterday, and – apparently on their own initiative – painted neat black and yellow stripes along the edges. We pass another group of teenagers who are setting out with brooms to tidy up a city square. 'We're all volunteers now,' says Molly. 'There's so much energy. We all know the challenges but . . . there's such a buzz here.'

Molly's helping out in what passes for the local administration, translating documents, writing letters, trying to build a brand-new state from scratch. 'It's chaos,' she admits. 'Some days I think we'll never get anywhere. But we are making progress. Perhaps it's naive, delusional, but we believe we can win this war.'

I pass a tiny stationery shop, and its owner, eighty-year-old Abu Salan Tashani, rushes out to say hello. In 1947 he opened a book-shop just down the road. 'But I sold foreign books, history books – Gaddafi didn't like them,' he says. 'The police came and gave my shop to someone else. I never got it back.' Mr Tashani takes out a pile of fading souvenir postcards, showing Benghazi in the 1970s. I ask him if Gaddafi's time is running out. He shrugs and raises his eyebrows hopefully.

A burst of gunfire echoes off the walls. I'm getting used to it now. The shooting normally starts in the evening: a casual, festive chorus directed mostly at the sky. One night, some boys took playful, wobbly, aim at the lagoon outside our hotel. On another occasion, a dozen massive explosions turned out to be dynamite, thrown into the same placid water for fun.

It's surprising the city feels so calm, given how many guns and grudges there must be. NATO fighter jets can take some of the credit for keeping Gaddafi's army at bay. But the tribal and political rivalries that still cause so much concern across Libya seem almost invisible on the streets here.

After a day visiting the deadlocked front lines an hour and a half south of Benghazi, I arrange to meet a man called Muhammad Abu Nnaga. In 1998, he and some friends were arrested by Gaddafi's police, taken to Tripoli, beaten, starved, and then held in a tiny dark cell for three years. Only then were they charged – with plotting a coup.

'It was ridiculous, we weren't political,' says Muhammad. 'I was a medical student. We were interested in charity and civil society. But we were organised, and the regime saw that as a threat.' The judge turned out to be a distant relative. But he still had his instructions, and sentenced Muhammad and nine friends to ten years in prison.

Muhammad was released early in 2006, at the age of twenty-nine. 'Gaddafi was trying to make peace with the West,' he says. 'I came home scared. My parents were scared too. I felt dead – as if there was no more meaning to life.'

As he tells me this, on a bench overlooking the harbour, the crowds start to gather nearby, ready for another afternoon of speeches, marches and revolutionary songs. 'Now I think we're in for a long fight,' says Muhammad, as he heads off to join a youth group he's organising. 'Gaddafi's greatest crime was to take away our dreams. But now we have our dreams back, and we won't let them go.'

Allan Little – Tunisia – Election Day at Last – 27/10/11

The first election to emerge from the Arab Spring was held in Tunisia amid high hopes. President Zine El Azedine Ben Ali, in charge since 1987, had been forced out by a month of intense street protest and fled for Saudi Arabia on 14 January 2011. After an edgy interval as legislators and politicians scrambled to arrange the first truly free elections in a generation, the polls finally opened in October. The centrist-Islamist party, Ennahda, which came out ahead with 41 per cent of the national vote, announced it had started coalition talks to form a new government. Allan Little noted that while some (both inside and outside the country) were disappointed that an Islamist bloc did so well, there was hope that a genuinely multi-party, secular democracy would emerge.

In the courtyard of a secondary school in central Tunis the sun was pounding down by 10 a.m. The queue snaked one way then doubled back and then twisted again, filling the space with patient, cheerful people. 'I've been standing here in the sun for three hours,' one man said. 'We are suffering here, but you know, it's a beautiful kind of suffering.' How long have you been waiting to vote, I asked? The answer was irresistible. 'Forty years,' someone said.

'Who do you think will win the election?' I asked another voter. 'Sir,' he said, 'we will *all* win. We are voting freely for the first time. It means we have *already* won'.

There was a generosity to this day too. People who'd waited hours happily stepped aside to allow the elderly to jump the queue, even though this slowed everyone else down.

But when a prominent party leader tried to do the same thing, Tunisians flexed their newly acquired democratic muscle. Rached Ghannouchi, the leader of the moderate Islamist party Ennahda, turned up at his polling station surrounded by television crews and made straight for the door. 'Hey!' someone shouted. 'Where do you think you're going? Can't you see there's a queue?' Soon the rest of those waiting joined in. Chastened, the modest Mr Ghannouchi took his place in line and waited like everyone else.

There aren't many days in a life spent chasing news that are as unremittingly positive as this. The days that remind you that the right to vote is held most precious by those who have long been denied it and who are exercising it for the first time.

There was something pleasingly fresh about the campaign, too. There were no spin doctors, no rebuttal units, no focus groups, no electoral machines to harvest votes. The glib and oily arts of the more mature democracies have yet to take root here. It was politics on the pavement: Tunisians rejoicing in the right to disagree in public, to press a leaflet into the hands of a neighbour in a shopping centre, to post a flyer on a billboard. Ten months ago, they could have been jailed for doing any of these things. Still, more than a hundred registered parties emerged to contest the election. Crucially, none of them is backed by an armed militia.

The campaign really has been about argument, about competing visions of the country's future. And the conservative vision prevailed. Mr Ghannouchi, who was sent to the back of the queue by his neighbours, emerged as the most powerful man in the country.

Why have Tunisians chosen the Islamists? I sat in a café in a working-class neighbourhood where unemployment is high and disillusionment and disgust with the old dictatorship are deepest. Ask people here why they voted Ennahda and they don't talk about religion. They talk about honesty in public life, about the need for a government that won't steal from the people. This is the reputation that Ennahda brought to the electoral table. They won not because people want an Islamic state, but because people think they represent the best chance of a clean break with the corruption and venality of the old regime.

In the smarter, wealthier, less socially conservative neighbourhoods, where almost no one wears the Islamic headscarf, young people hang out in pavement cafés in streets that could easily pass for suburbs of Nice or Marseilles. Here, there is unease but not widespread alarm. No one expects that women will suddenly be required to cover up, or forced from the workplace. In fact, gender equality in the workplace, including equal pay for equal work, is one of the policies of the Islamist party.

'Are you afraid of an Islamist dictatorship emerging here?' I asked the leader of one secular, extremely liberal party. 'No,' he said. 'Don't think of Tunisia as a battle between Islam and democracy. Think of it as a battle within a democracy between two competing visions for our future – one is conservative, rooted in religious observance; the other is progressive, modern and forward-looking.'

Tunisia has a strong, educated, self-confident middle class that, it is now clear, has a very developed understanding of the dynamics of democratic life. This country is in the cockpit of something new in the world: a real, indigenously rooted Arab democracy. Tunisians toppled their dictatorship. Tunisians pulled off this remarkable election. Tunisians own their revolution. They have embarked on a future of their own choosing.

Yes, these are early days. Yes, a lot can still go wrong. But right here, right now, this feels like a moment of real consequences, and great promise.

Lyse Doucet – Egypt – The Sofa Overlooking Tahrir – 26/11/11

Sometimes the best way to tell a vast story depends on finding the right vantage point. In the run-up to the promised parliamentary elections after Mubarak's fall, the political temperature soared on the streets of Egypt. Secular, middle-class reformers, left-wing groups and feminists as well as the long-suppressed Islamists bridled under the temporary control of the Supreme Council of the Armed Forces. There was suspicion the army wouldn't let the elections happen. Protesters fought in the streets with army and police, and dozens were killed. Lyse Doucet witnessed the turmoil in Cairo.

Manal sits in her chair most of the day, much of the night. It's a big stuffed chair, covered in crushed velvet, patterned with flowers. She's a big Egyptian woman, in a flowing robe and a headscarf that frames a round, kind face. From her chair, Manal looks across her bedroom through open balcony doors. And if she tires of looking there, she shifts her gaze to a little television set.

But she doesn't need to watch TV these days. Manal has a window right onto what the world is watching now – Tahrir Square. From her apartment above this now-famous space, she hears it all: running battles between police and protesters, ambulance sirens wailing long into the night. Sometimes, the tear gas fired by police is so strong it wafts through her windows. Doesn't it keep her awake? Manal smiles sweetly, eyes crinkling behind wire-rimmed glasses. I can't sleep, she says, for worrying about these young people in the square. I need to know they are safe.

And what about the sons of this widowed mother? Fourteen-year-old Ismat listens shyly, clutching his school books. He's only allowed to go down the five flights of stairs to visit the field hospital right at their door. Hundreds of protesters are treated there every day. Manal's older son Ahmed has more freedom. He crosses the square to meet his tutor on the other side, and then crosses back to come home. Does Manal worry? Ahmed, she says, bids her goodbye with an expression often used by pious Muslims: 'If I have life before me, I will return.' In other words, it's in God's hands.

Ahmed sees a Tahrir Square his mother can't see from on high. On the ground, it is in constant flux, becoming a city within the city. Young men and women form a human cordon checking all who enter. Inside, volunteers clean the streets, doctors in white coats treat the injured, and enterprising street vendors turn a profit. When hundreds of thousands of Egyptians recently converged on the square, there were flags and trinkets of the revolution for sale along with the red toffee apples and pink cotton candy. Now you can buy everything from roasted sweet potatoes and salted nuts to gas masks, sunglasses and socks. It can feel like a carnival, but more often these days it's a confrontation.

At night, from Manal's balcony, it's a seething human canvas – a carpet of yellow and white light threaded with flashing blue ambulance signals … you see crowds surging towards police and army lines, then running back to escape the tear gas. Human chains shout slogans like *'Erhal! Erhal!'* 'Leave! Leave!' – that's what they chanted when they first took over this square in January and called on President Mubarak to go.

Now they're back in the square, using the same slogans to demand that the real power in the land, military chief Field Marshal Tantawi, must go. And this time, activists say, they're not leaving until all their demands are met. Many well-known activists traipsed up the five flights of stairs to Manal's balcony this week. They looked out with palpable awe and anger: awe that Tahrir Square is throbbing and thriving again; anger at how many people have been killed and injured by security forces.

Tahrir has found its voice again, and Egypt's military feels it must listen … but only up to a point. On the crowded streets of Cairo, I met many Egyptians who thought Tahrir's time was gone, who believed the army was doing what it could to move Egypt towards democracy, including calling elections. When we visited villages along the Nile, we found the tumult in Cairo had not escaped them, but life, with all its growing hardship, still has to go on.

For Egyptians like Manal, Tahrir has taken over her life, disrupted by the tide of people, including journalists like us, with whom she's kindly agreed to share her view. For now, she watches serenely from her chair, unable to say when and how it will end. No one can.

Gabriel Gatehouse – Libya – *The Gun Is My Ministry* – 18/02/12

A year on from the start of the revolution there were celebrations in towns and cities across the country, but any rebuilding of Libya as a state was being hindered by ever-multiplying armed militias. Amnesty International claimed that they'd committed widespread abuses of human rights, among them war crimes against suspected Gaddafi loyalists, unlawful detention and cases of torture, sometimes leading to deaths. After decades of iron rule from Tripoli, the situation was almost anarchic, with no central authority able to impose order. Gabriel Gatehouse had several run-ins with the gunmen who were starting to carve out their own spheres of influence by force – or threat – of arms.

Benghazi is a city on edge. The flood of adrenalin that powered the revolution has ebbed away, exposing frayed tempers and short fuses.

We went to the hospital to interview a doctor. During the revolt against Gaddafi he had, for a brief period, swapped his stethoscope for a gun. The doctor's name was Ahmed el Metjawel. He met us at the main entrance and we hadn't got very far when some former rebel fighters, acting now as security guards in the hospital, asked us if we had permission to film.

The doctor said he would answer for us and on we went. But before we could reach his office, the same security guards reappeared, running after us, with some more senior doctors in tow. An argument broke out between our friend and one of the hospital directors. The row was conducted in more or less civil tones, but resentment bubbled under the surface.

Eventually we made it to Dr Metjawel's office and started recording our interview. We had barely begun when those same three fighters burst in and told us to stop. Now the anger boiled over. The doctor and the security man squared up to each other in the tiny office, nose to nose.

We needed permission from the commander of the local brigade, shouted the fighter. What right did he have to walk into his office and tell him what to do? responded our doctor. And so it went on, hands gesturing, arms flailing. Eventually the security man threatened to call in the *katiba*, the men with the guns.

In Libya, the gun is the ultimate arbiter of disputes. It's not always used, just as it wasn't in this case; sometimes the mere threat is enough.

A few weeks ago I went to check on reports that there was digging going on in Colonel Gaddafi's former compound in Tripoli. The dictator's gold, it is rumoured, lies hidden somewhere in the ground. A group of men stopped me from entering. On whose authority, I wanted to know?

'Authority?' the man in charge asked.

'Yes,' I said, 'which ministry or government department has decreed that the BBC should be prevented from seeing your digging?'

'I don't need any ministry,' said the man, patting his Kalashnikov. 'This is my ministry.'

The incident at the hospital in Benghazi was eventually resolved. But that little argument told us a lot about what kind of a country Libya is becoming. It is a place where the *katiba* rules supreme. This country is run by a patchwork of former rebel fighting brigades. Like the *katiba* at the hospital, most of them are answerable to no one but themselves. Each rules over its own little territory, whether it's a hospital, an airport or a few city blocks. It is at the edges of these small fiefdoms that the trouble starts. If ever there were a place that proved the old adage that all politics is local, then Libya is it.

I had wanted to ask Dr El Metjawel, who'd fought in the revolution, how he felt about the direction Libya was heading. But his main concern seemed to be the management of the hospital: it was the same old Gaddafi-era figures in charge, he complained. And they were simply appointing their own family and friends to positions of responsibility.

The same local concerns had upset the security people. They didn't *really* mind us filming in the hospital. But we had asked the doctor for permission rather than the fighters, undermining their little patch of authority.

Almost every evening you can hear evidence of these minor disputes as the sound of gunfire drifts across Libya's cities after dark. And yet a recent survey suggests a staggering degree of optimism amongst people here. Whatever their frustrations about how things are turning out, more than 90 per cent of those questioned thought the revolution was a positive development. And so the Libyans are both united and at the same time dangerously fractious.

The day after that incident at the hospital, I sat in the morning sunshine talking to a man who has become something of a professional protester. Every day he comes out onto one of Benghazi's main squares to voice his displeasure at the failings of the local government, the power of the armed militias. After I'd listened to his litany of complaints, I asked him if he thought it had all been worth it. His face changed. 'Oh,' he said, 'I feel like a different man now. I can breathe. I am free.'

Shaimaa Khalil – Egypt – My Father, the Nation and the Army – 27/04/13

While Egypt's revolution had promised democracy, before long the messy reality of life under the new president, the Muslim Brotherhood's Mohamed Morsi, led many to question whether military rule might have had its good points. Shaimaa Khalil offered personal insight into how, barely two years after overthrowing Mubarak, the nation might be willing to hand power back to another man in uniform.

'It was always my dream to be in the army,' Lieutenant Colonel

Mostafa Salah El Dine told me with a ring of pride in his voice. He's tall, dark and in his mid-forties; despite having left the armed forces almost ten years ago, he still has that authoritative army walk.

'When I was young, I and most of my friends wanted to become army officers,' he told me. 'When you're in the army, they really teach you how to love Egypt . . .'

This proud tone is quite familiar to me. I heard it many times among my military family. Especially when my grandfather spoke of his memories. 'You see this?' Grandpa would tell me, pointing proudly at shrapnel on the ring finger of his right hand. 'I got this in the war!' He wouldn't say which one.

Growing up in Egypt, you are taught from a very early age that the armed forces are the defenders of the nation – 'Allah's soldiers on earth', as they are commonly called in Egypt, giving them that extra bit of divine credibility.

My earliest memories are of my father picking me and my sister up from school. He'd show up in his pressed olive uniform and his beret and the all-important epaulettes with a golden star and eagle, worn by lieutenant colonels like him. I remember how much respect he commanded in our neighbourhood; he was well liked, but I'm sure the uniform helped.

He was a man of very few words, and when he spoke he was assertive and firm. No negotiations on our curfew and no excuses if I was late, even for a few minutes. The perfect candidate for the army, you could argue. Photographs of my father's graduation from military school and hand-painted portraits of him and my grandfather in full uniform have always been on display in my grandparents' front room.

There's nowhere that pride is better displayed than at the entrance of Cairo's 1973 War Museum. As I walk in, I'm surrounded by old MiG fighter jets, tanks, various rockets and guns. The rest of the world doesn't see this conflict as an Egyptian victory at all, but in Egypt there's no doubting that the army was triumphant in 1973. It's one of the main reasons why the military is held in such high regard here. Inside the museum, a film relays the stories of the Egyptian army's heroism. It's filled with patriotic songs which go down very well with local schoolboys, who burst into claps and whistles.

Egyptian troops were also once seen as the saviours of the 2011 revolution, when they sided with the people and helped oust President Hosni Mubarak. But that image changed when they took charge of the country. In the year and a half after the revolution, the military council's handling of day-to-day politics and their heavy-handedness in dealing with clashes changed their image from that of saviours to oppressors.

Egyptians' relationship with the military has always been complex, a mixture of respect and fear – and after the events of the last two years, outright resentment at times. 'We love the military, we respect them,' people will tell you, but then they say, '*Khalas!* – that's it! - sixty years of military rule is enough!' But when the security situation deteriorates – and in Egypt that's been the case for two years now – the same people will expect the army to save the country.

As a child I didn't know much about the army's vast economic ventures – the factories making everything from pasta to stoves, fridges to kitchen utensils – but I did know about its beach clubs. The one in Alexandria was not lavish, but it was good enough for me and my sister. Many a day was filled playing around in the sand, swimming and watching my father as he was fishing. The army also had housing projects for officers – big apartment blocks in one of the city's prime areas. When my father had his first heart attack, he was treated in Alexandria's biggest military hospital. It was the same hospital that he died in years later.

For my father the army was a constant presence in his life, from birth until death. And for Egypt, the influence of the army goes far beyond borders and battlefields. It runs through the veins of everyday life here.

Quentin Sommerville – Egypt – Bloody Monday – 13/07/13

Egypt's upheaval continued. An ostensibly independent grassroots campaign called Tamarrod began to gather signatures to remove

Mohamed Morsi from the presidency by the end of June 2013. There is still speculation over how far the campaign may have colluded with the Egyptian military. Huge (and still disputed) numbers of people took to the streets again near the end of June calling for Morsi to go; anonymous military sources claimed crowds of up to fourteen million across the country. In a twist of history which wasn't so much a backlash as a whiplash, after days of high tension, the army retook power on 3 July 2013, stripping Morsi of the presidency and arresting many other senior figures in the Muslim Brotherhood. The street encampments of MB loyalists who turned out to support Morsi met with brutal treatment. Quentin Sommerville saw the confrontation playing out.

The old woman in the floral headscarf had been standing in the sun for hours. Her arms were draped through the barbed wire of a barricade, just a couple of blocks from where, earlier that morning on a Cairo street, fifty people had been shot dead by the Egyptian army.

A distance away, there were hundreds of protesters behind her, mostly Muslim Brotherhood members and all Morsi supporters. Beyond the barricade were ranks of soldiers, armed with automatic weapons, tear-gas launchers and black gas masks.

Her name, she told me later, was Umm Ibrahim – a nickname meaning 'the mother of Ibrahim.' She wasn't easily intimidated. It was clear, even from a distance, that she was giving two of the soldiers, who were wearing body armour, a serious ticking-off. 'I told them,' she said, 'these people are your brothers, you're all from the same country, how can you kill another Muslim, while he is praying?'

She'd been down the street at dawn that morning, praying with thousands of others, outside the Republican Guard barracks. She, like the others, believed President Morsi was being held inside. As they finished prayers, she explained, the army opened fire. 'There was a lot of gas,' she said. 'Forty-five people were killed in front of my eyes.' How did the soldiers respond? I asked. 'They told me they were just following orders and that they would be shot and killed if they didn't.' And with that, she stomped back to the protesters, shouting at them to stop throwing stones, just for good measure.

The army has a different account of events. That they were attacked by protesters trying to get into the barracks, some of whom were armed. Dark forces were at work: Syrians, perhaps Palestinians were involved too. But eyewitnesses, even those who're against President Morsi, dispute the army version; and this is the same army that just days earlier opened fire directly into another crowd at the same spot, killing at least three. While the army is armed to the teeth, the protesters generally are not.

Aside from the killings, the most shocking thing was the lack of interest from anti-Morsi campaigners. I spoke to someone from one of the opposition parties. He said the reports were Muslim Brotherhood lies, and chose instead to focus on some viral pictures of dead children, purportedly from the scene, that had been exposed as fakes. Yes, the Muslim Brotherhood should take part in elections, he said, but in the same breath denounced the party as untrustworthy. 'They should stop inciting violence,' he added.

In the eyes of those who stand against the former president, the army are heroes who delivered a people's revolution. In Tahrir Square, a giant laser projector beamed the words THIS IS NOT A COUP onto the side of a building. A civilian interim president, Adly Mansour, is meant to be proof of this. But the army is still on the streets, and still issues statements telling political parties to get in line ... or else.

Millions did take to the streets to get rid of Morsi, who failed to unite the country, never managed to get a proper grip on the levers of state, and let security and the economy deteriorate under his rule. This, say his detractors, means that his democratic election can be annulled. And anyway, so the argument goes, at that election, the choice was between Morsi and the old regime, which was really no choice at all.

Whether this is a coup or a revolution doesn't really make much difference. It certainly isn't democracy, and it means plenty more trouble for Egypt. Because the Muslim Brotherhood and their millions aren't going away.

Back behind the barricades, pools of blood were baking in the afternoon sun. Occasional shots would ring out from the direction of the army. They were only shots warning people to stay back, but unnerving nevertheless. Adding to the surreal air, above, an apartment was ablaze. The fire filled the neighbourhood with smoke; occasionally blackened papers would drift past. But no one paid any attention; the fire was ignored. The focus was on the barricades and the soldiers.

A young man from the Brotherhood came up to me. He didn't give his name, but relayed a threat. 'We've managed to keep the jihadists under control,' he said, 'but only until now. They will pick up their weapons again, and how can we stop them after what happened at the barracks?'

The army said they intervened in politics to unify a divided society. President Morsi had to go because he didn't represent the entire country – religious and secular, rich and poor. The tragedy for Egypt is that neither does his successor.

Charlotte Bailey – Tunisia – Still Burning – 21/07/18

In Tunisia, too, came disillusion. The economy stalled, especially after terror attacks in Tunis and Sousse in 2015 damaged the tourist sector. Inflation rose, unemployment stayed high and the foreign investment that was meant to flow into the country after Ben Ali's corrupt regime ended failed to materialise. For young people especially, the prospects still looked bleak; thousands tried to leave for Europe, while others made their frustrations known in other ways. Charlotte Bailey found one particularly painful echo of the recent past.

As I walk down the corridors of Ben Arous Burn and Trauma Centre in the south of Tunis, I can hear the patients screaming.

This is the only specialist burn unit in the country, the place where the fruit-seller Mohamed Bouazizi was brought in December 2010, after he set himself alight in response to harassment by city officials. For the doctor who treated him, Amen Abdel Messadi, who's talking to me between appointments, Bouazizi was, in many ways, just another patient. But what Dr Messadi calls 'the noise' around him was quite different from anything he'd experienced before.

As Bouazizi lay in his hospital bed, street protests grew. Tunisia's then president, Zine-al Abidine Ben Ali, even came to visit, in an effort to turn around public opinion. But the patient died on 4 January 2011; the president was ousted ten days later. The fruit-seller had sparked the Arab Spring. But his act didn't just start a wave of protests that would topple governments in Egypt, Libya and Yemen. It also triggered a disturbing wave of copycat self-immolations.

'Of course,' Dr Messadi tells me authoritatively, 'self-immolation didn't start with Bouazizi. It is a cause of burns all over the world.' There was, however, a dramatic rise of cases after Bouazizi's death.

Everyone expected the copycat immolations to be a short-term phenomenon. But doctors in Tunis are horrified that, more than seven years later, they continue at the same disturbingly high rate. Dr Messadi flips through his records, frowns and shakes his head. 'There were three hundred and twenty-five patients admitted here last year. Of those, eighty-eight did it to themselves.'

It is a brutal way to die. 'It doesn't kill fast. It is slow and very, very painful,' Dr Mehdi Ben Kheilil, a forensic pathologist at Charles Nichol Hospital in Tunis, had told me the previous day. For medical staff to have people rushed through the doors who have deliberately set fire to themselves, day after day, month after month, year after year, is heart-breaking.

'The people who do this are very young. In general, they are under thirty,' Dr Messadi says. Mostly, they are unemployed men.

The protesters who took to the streets in 2010 were demanding not only democracy, but jobs. While some of the political demands have been met, unemployment not only persists in Tunisia, it has worsened. The national unemployment rate is 15 per cent, but it's a staggering 35 per cent for under twenty-fives.

Most of the men set themselves on fire in front of public buildings. Psychiatrist Fatma Charfi who runs the suicide prevention committee at the Ministry of Health told me, 'It is an act of social protest and a way of transmitting a message. In this way there is sacrificial symbolism to the suicide.'

Dr Mehdi Ben Kheilil agrees: 'They want to make a display of how they are feeling.' He is a warm, calm presence, but tenses slightly as we turn to the subject of the reasons behind the self-immolations. He blames economic factors, but also journalists, particularly the national media, which, in the immediate aftermath of revolution, he says, idealised self-immolators and held them up as martyrs.

Salem Bouazizi blames the government not only for his brother Mohammad's death but also for the copycats. He had harsh words for politicians: 'Stop your hypocrisy and your plotting for power and pay attention to the country and its people – especially the youth.'

Whatever drives young men to set themselves alight, their families are left bereft. Salem told me he couldn't bear to remember the pain he and his family had been through over the past years. 'May God have mercy on his soul,' he said.

But not everyone dies: the survival rate is around 50 per cent. Survivors are doomed to live on, in an even more difficult situation than before. They are left terribly scarred; the skin around the mouth often retracts so they cannot speak, and they are frequently left unable to move their limbs. The treatment for severe burns costs a lot of money and takes up to two years – a real problem, given that most of the people who do this are impoverished to begin with.

Families often resent the burden. Survivors may be rejected by society because of the burns' appearance, but many religious Tunisians also consider suicide, or attempted suicide, to be blasphemous, the doctors tell me. 'Society is very hard on these people,' says Dr Messadi. 'People think, well, you did this to yourself.'

He has just finished presenting the results of a survey into the lives of surviving self-immolators. 'They all say that life has become much harder.' He says he sees it every time a patient comes around: 'They regret, they all regret, as soon as they realise what they have done . . . but, of course, by then, it's too late.'

Kevin Connolly – Israel – The Lemon Tree on My Balcony – 21/01/12

International reporting on the conflict between Israelis and Palestinians returns almost obsessively to questions of land – not just over who owns a particular spot, but also over the rights to live, build and farm on it. This perspective sometimes misses another resource which might be even more important: water. Since the Six Day War of 1967, Israel has had access to the water of the West Bank of the River Jordan – and ever since, there have been arguments over how it should be shared and managed. While living and working in Jerusalem, Kevin Connolly found that even his back yard could tap into far wider questions.

There is a quiet satisfaction in the patient husbandry of traditional gardening – that gentle and miraculous coaxing of new life from the unpromising earth. But there is a rapid and more convenient satisfaction to buying your shrubs and trees already fully grown. Watching our own lemon tree being manoeuvred off the back of a flatbed truck by two burly delivery men took half an hour, rather than half a lifetime. There is an economic issue, admittedly: so far, the tree has yielded four lemons which must have cost about £100 each, but I'm hoping that the average price will fall over time.

For all the fragile grace with which it dances in the chilly winter winds, the tree conceals an astonishing number of thorns beneath its gentle leaves – although at the moment it's pricking my conscience more than my fingers.

The problem is that like most balcony gardeners in Israel, we have installed a miniature irrigation system to keep the tree alive in the brutal heat of summer. It doesn't amount to much more than a couple of metres of plastic piping and a timer attached to a tap, but every time I hear the muted sloshing of another carefully calibrated dose, the desert around us feels a little drier.

Often the water issue here is reported as part of the broader tension between Israelis and Palestinians. A French parliamentary report recently concluded that the 450,000 Israeli settlers who live on the West Bank of the River Jordan, in defiance of international opinion, use more water than the 2.3 million Palestinians whose home it is.

The fairness – or lack of fairness – with which resources are shared is important, of course. But there is a larger issue which may in the end be more important still. That is the alarming way in which the amount of water in the lakes and rivers here is dwindling.

The River Jordan carries water south from the Sea of Galilee to the Dead Sea, passing as it does through Palestinian, Jordanian and Israeli land. These days the Jordan in many places is hardly more than a listless and polluted dribble, but there is plenty of evidence that it was once very different. Halfway along the valley is a hydro-electric power plant, long since abandoned. It's a sobering thought that once the river waters turned its mighty turbines, when these days they are hardly potent enough to moisten a handkerchief.

That's partly because these days Israel pumps water out of the Sea of Galilee to feed its national supply system and partly because neighbouring Arab countries use water from the rivers that feed the Galilee. The effect on the River Jordan is measured best by watching what is happening to the level of the Dead Sea into which it flows.

It remains one of the natural wonders of the world: the lowest point below sea level on the surface of the planet, with its own sweating, sulphurous microclimate, and water so dense and salty that you can neither sink nor swim in it. But the most striking point about the Dead Sea is that it's shrinking – by a metre's depth a year. It's now only about two-thirds the size it was in the 1930s.

The ancients once believed it was certain death to try to sail across the Dead Sea; give it another hundred years or so and you might be able to step over it. So something has to be done – and attitudes to water in the Middle East are not always rational.

I wince when the lawn-sprinkler system at my apartment complex switches itself on. Then there are the televised golf tournaments – from elsewhere in the desert but played on the lushest of grass. Bunkers and sand traps, not putting greens, mimic the Middle Eastern terrain, so God knows how much that defiance of natural circumstance must cost.

In its modest way our lemon tree is not helping either; every time I hear the gurgle of the irrigation pipe I imagine the Dead Sea shrinking a little further. By the time we come to leave Israel, I'm told the tree will be too big to fit in our building's lift, so a crane will have to be hired to winch it off the balcony – thus probably raising the average price of the lemons again. Instead of selling it, I'm tempted to take it down to the shores of the Dead Sea and replant it there. Deprived of its artificial life-support system, it might not thrive – but perhaps it'll do it good to learn to fend for itself.

Turkey

Chris Morris – Is Turkey Getting Hairier? – 22/06/13

Since becoming Turkey's prime minister in 2003, Recep Tayyep Erdogan proved an adept political operator with a talent for giving voters what they wanted. His background as mayor of Istanbul seemed to show he could get things done, while sustained economic growth reassured people the nation was safe in his hands. On hot-button issues – from dealing with the country's Kurdish minority and to the role of the military, to coming to terms with the mass killings of Armenians nearly a century earlier – he had fine-tuned a message combining pragmatism and national pride. Still, there were rumblings of discontent about cronyism, dubious state contracts and his hard-driving, pious, autocratic style. Things came to a head in the summer of 2013 when his plans for a huge redevelopment of an area of central Istanbul sparked mass protests. Over 2.5 million Turks took part, and the hashtag #OccupyGezi rallied support from around the world. But Chris Morris cautioned against assuming that secularist, modernising, internationalist democrats would always carry the day.

It's been a very modern sort of protest, fuelled by street art and social media, deploying irony and humour to prick the pomposity of the powerful. Prime Minister Recep Tayyip Erdogan has called anyone who drinks alcohol an alcoholic. So protesters toast him with beer bottles in the street. He's advised Turkish women to have at least three children. So protesters wear T-shirts which gently mock him. 'Tayyip,' they ask, 'do you really want three of ME?'

From the Whirling Dervish spinning in a gas mask to the Standing Man of Taksim Square, everyone has been out to make a

statement. The solitary protest of Erdem Gunduz – standing silently for eight hours in an area surrounded by riot police – has inspired
hundreds of imitators. 'We think in a different way,' he told me in a
back-street bar at three in the morning, shortly after the police had
brought his personal protest to an end. 'The government finds that
rather hard to accept.'

And yet for all their inventiveness and determination, these
protesters represent a minority – a sizeable one, but a minority
nonetheless. Millions and millions of Turks live in a rather different, more conservative world, and the past decade has been good
to them.

There's no doubt that the country has developed and changed
dramatically since I lived here at the turn of the century. Better
roads, better hospitals, more wealth in the provinces – the list goes
on. But it's often the small signs that make the biggest impression.

Like the guy who was manning a Turkish Airlines check-in
counter, sporting a little bushy moustache. In the complex code
of hirsute Turkish politics, that means Islamist. He wouldn't have
been sitting there looking like that a decade ago, in a more secular
Turkey. 'The nation is becoming more hairy,' said one of my friends,
with a half-raised eyebrow. And why not?

Want to know why Tayyip Erdogan has won three successive
elections? Mainly because he's spread prosperity, but also because
many of his supporters feel he's liberated them from decades of
what amounted to secular fundamentalism. Women wearing headscarves were banned from university; men with beards were treated
with official suspicion. No longer. The trouble is that Turkey needs
a delicate balance, it always has done, and the impression is growing
that Tayyip Erdogan is now swinging the pendulum firmly in the
other direction.

Recent comments made by an influential official in Mr Erdogan's
AK Party are revealing. 'Liberals who supported the party over the
past ten years,' he said, 'will not be with us over the next ten years.
They will not like what we intend to build.' This new construction
appears to include more and more restrictions on drinking in any
public place, no kissing on the subway, and turning old buildings,

parks and sometimes entire neighbourhoods into a neo-Ottoman Disneyland, glorifying the Islamic past.

When people complain, it's Tayyip Erdogan who leads the charge against them. 'Why,' the prime minister asked recently, 'are the laws crafted by two drunkards respectable, while the laws dictated by religion are rejected?' That was a none-too-oblique reference to the first two presidents of the republic, the iconic Ataturk and his successor Ismet Inonu. And for many people that crossed a red line – by insulting Ataturk and suggesting the laws of Islam should help run the country, all in one sentence.

So now there's impasse, and old fault lines are being re-examined. Not just between Islamist and secular, but also between right and left, and – worryingly – the sectarian line between Sunni Muslims and the minority Alevis. The prime minister rejects accusations that he seeks to impose the tyranny of the majority and highlights his bold effort to seek peace with Kurdish separatists to prove his point. But much of his recent rhetoric about the protests has been divisive: it's all a vast conspiracy; you're either with us or against us.

Is there anything outsiders can do but watch and worry? Well, for much of the past twenty years the carrot of EU membership has helped drive reform in Turkey. It's been a process fraught with difficulty. Turkey may well never join, and may choose not to do so. But now there are those – in Germany in particular – who seek to shut Turkey out for good. It seems to me that that would be a big mistake. Huge numbers of Turks have spent the last few weeks out on the streets, showing how much they value European ideas of pluralistic democracy. And sometimes it's the symbolism that matters.

Mark Lowen – Coup Aftermath – 20/08/16

As prime minister, Recep Tayyep Erdogan led the AK Party to three general election victories in succession. Having outlasted the Gezi Park protests, his control of Turkey tightened even further when he became the country's twelfth modern president in August 2014. His critics might have been less visible on the streets, but there certainly were still people trying to remove him from power. Yet their efforts backfired. After some elements in the armed forces attempted a coup d'état on the night of 15 July 2016, even many of the president's former enemies jumped to defend their democracy and his leadership – and to denounce anyone or anything they considered a threat. Mark Lowen explained how the failed coup ended up consolidating Erdogan's position.

He'd been at the rallies in Ankara every night, with his elderly father, selling Ottoman-style fez hats. His smile and eyes were full of pride as he talked of his love for the president. His name, appropriately, was Ali Erdogan. On the night of the failed coup, he responded to President Erdogan's call to flood onto the streets and stop the tanks. And then, for over three weeks after the takeover was crushed, he joined tens of thousands at nightly vigils in what they called a 'democracy watch'.

Rolling up his sleeve to show his shrapnel wound, Ali told me he'd do it again. 'May my last drop of blood be given to this country for our flag and our people. We will always stand with Turkey and with our president,' he said, before turning to negotiate another sale.

Suddenly a woman approached, shouting in English: 'Why are you interviewing him, the fez-seller – it's a cliché! Why are you not filming the crowds to show the unity of Turks? You Western media are all the same.' She broke out in a chant: 'We love you, Erdogan! We love you, Erdogan!'

Two sentiments that have been so strong during this most tumultuous month for Turkey: an outpouring of nationalism and fury against Western governments. Two hundred and forty people were killed resisting the military coup in which parliament was bombed and the president almost captured. Tanks drove into or fired at those in their way. It's all prompted a frenzy of patriotism. Cross-party rallies have gathered millions; talk of political unity has been loud. But so has anger against allies who many here believe have abandoned them.

Western leaders quickly condemned the attempted coup – but

none has come here to show support and their statements have focused more on the tens of thousands arrested or dismissed in the biggest purge in Turkey's modern history, than on the coup itself. There's also the thorny issue of the alleged coup mastermind, the cleric Fethullah Gulen, in self-imposed exile in the USA since 1999. Ankara is demanding his extradition; Washington says it needs proof of his involvement. The row has poured fuel on the anti-Western fire, with headlines screaming: 'America – choose Turkey or the terrorist!' and polls showing vast numbers of Turks believe the CIA was complicit in the coup attempt.

It's been years since relations were so poor with this country, an ally the West badly needs. Turkey has the second-largest army in NATO; it's still negotiating EU membership – although that's a goal that may never be reached. After Ukraine's revolution in 2014, EU and NATO chiefs flew to Kyiv to stand where dozens had died in Maidan Square. There has been no such display of solidarity in Turkey. In the eyes of many here, anti-Erdogan feeling in the West has trumped support for democratic resistance to the coup. From the perspective of critics abroad – and those still here – an authoritarian president is seizing the opportunity to tighten his grip.

Amidst the sea of flags and the chorus of nationalist songs at the rallies, another passionate feeling becomes clear: that for vast numbers of Turks, the night of the coup attempt has become a moment of rebirth. Banners commemorating the martyrs and the nation's bravery hang everywhere. Even Turkish Airlines has a special on-screen message of praise for Turks' defence of democracy and the 'Commander-in-Chief, President Recep Tayyip Erdogan'.

For him and his supporters, this is their moment. Since the Turkish Republic was formed in 1923, the national hero has been its founding father, Ataturk, with big annual holidays linked to his achievements. That riled the conservative, pious side of the country that Mr Erdogan represents. He is their hero – the man who gave them a voice after decades of secularist rule. And now they have their own date – 15 July – when the Turkey where they're in the ascendancy came into its own, when Turks responded to the call of the commander-in-chief to stand against the tanks, when their own were 'martyred'.

A month on, a governing party MP was guiding me around the destroyed parts of parliament, craters gouged into the roof by the coup-plotters' rockets. We stopped amidst the shattered glass. 'July fifteenth brought us together,' she said. 'It was our liberation war.'

That is the narrative that has emerged in the past month: liberation from military coups, enshrining an era when Islamists no longer feel subservient to secularism. Half the country fears it. But the other half is savouring that freedom.

Syria

Lina Sinjab – Damascus Abuses – 24/07/10

Bashar al-Assad inherited the presidency of Syria by fiat: when his father, Hafiz al-Assad, died in 2000, the constitution was rapidly rewritten to allow Bashar to take up the post, even though he was six years under the official minimum age of forty. His leadership was then confirmed with a national vote ... which was reported as approving him by 99.7 per cent. Despite this start, there were signs of genuine reform during his first decade in power; some dared to talk about a 'Damascus Spring' in the early 2000s as political prisoners were released and the economy liberalised. But it was mostly window dressing, and the limited new freedoms were soon rolled back. Many argued that the old guard of Hafiz al-Assad's military-security regime was still in charge behind the scenes.

Driving through Damascus, there's plenty of evidence that Syria has changed during the last decade. Private banks are operating, the stock market is up and running, mobile phones and the internet are widely used. Women in designer headscarves, wearing designer sunglasses, totter around designer shops – while young people cluster outside the Western-style cafés, bars and restaurants which have sprung up all over town. But there's another layer beneath the glittering facade.

'They are controlling what the people will see, what the people will act like, and what they are allowed to do,' a young skateboarder told us, before checking that we wouldn't mention his name. We were on the windy slope at Tishmeen Park overlooking the city. Families were picnicking on the grass, with children playing football, shouting and screaming, and right above us towered a giant

Syrian national flag, blowing in the wind.

It had been raised to celebrate President Bashar al-Assad's ten years in power, a decade which other Syrians were celebrating. 'The ten years have been very good,' said one elderly man, as his daughter took photos of us. I asked if society had become any freer. 'No – it's *always* been free,' he replied rather stiffly.

Not far from the park, I squeezed onto a sofa inside an air-conditioned office, tense with expectation. A mobile phone rang. On the line was a human rights activist jailed on charges of weakening national morale. He was calling his family from prison. 'Someone like me should expect to go to prison – and should expect the battles to have consequences,' he told us on a crackly line, as we huddled round the speakerphone.

Not all his family know he is in prison. His mother, who wasn't with us, believes he is working abroad. In fact he's packed into a cell with seventy-five other prisoners, some convicted of murder and robbery. There are only fifty beds. He's one of those who have to sleep on the floor. 'It hurts me to be found guilty by the court,' he said, emotionally. 'More than the imprisonment itself,' he added, before another voice interrupted and the line went dead. The repetitive beep of the engaged tone was the only noise in the room.

Many human rights activists have been imprisoned over the years here. One of them, writer and journalist Fayez Sarah, has only just been released. 'It was really bad. I was under psychological pressure. I was beaten twice and put in solitary confinement for twenty-five days,' he told me. He was convicted of spreading false information about the state, but believes he was locked up because the government saw him as a threat. Sitting in his basement office, cluttered with books, with a small fan the only source of breeze to combat the oppressive heat, he seemed in good spirits and still hungry for change. 'I think in the coming time there will be political, economic and social development,' he tells me. Future generations here, he hopes, will have it better.

The government completely dismissed a recent Human Rights Watch report which criticised the state's treatment of people like Fayez. When we spoke to Deputy Prime Minister Abdullah

Dardari in his new marble-floored office, he maintained that it was impossible to take seriously every report about Syria published abroad. 'If you believed every word you read about this country,' he said, 'you'd think we were still in the Middle Ages', adding: 'I am not in the habit of commenting on what foreigners sitting abroad say about Syria.'

Instead, he praised the government's achievements, explaining that reforms had moved the country from a controlled to a free economy. 'Syria under President Bashar al-Assad has moved forward, has become a very dynamic civil society,' he tells me. Many people agree with that. Both young and old people can see the difference and, whenever we ask them, insist that they are proud of their country.

In the historical heart of the old city, in a small square garden filled with stone sculptures beside a cobbled road, I looked on as Damascene boys and girls mingled with international students who've come here to learn Arabic. For the young Syrians, a lot has changed. The simple fun of mixing with foreigners and voicing opinions was not widely available ten years ago. 'Bashar al-Assad is a young leader and knows what the youth wants,' one of the young people told me. Others said it was going to take time. The president, they said, had already made a lot of changes ... but he couldn't do it all at once.

Lyse Doucet – The Truth in Hidden Notes – 08/10/11

'The people want the fall of the regime!' The chant heard across North Africa was soon shouted in Syria too. By autumn 2011 tens of thousands of people were in the streets calling for President Bashar al-Assad to step down. The army shot back, using live rounds, tear gas and rubber bullets, often on unarmed demonstrators, sometimes elderly people, women and children. The fear of arrest and disappearance at the hands of the secret police was palpable. The UN estimated that nearly 3,000 Syrians had been killed since the pro-democracy protests began that spring. Lyse Doucet was allowed into the country on a journalist's

visa, but had to work while being closely monitored by government minders – and got a hint of how much the media were not being allowed to see.

It took a long time to get a visa for Syria and I will remember this trip for a long time, as well as holding on to some memorable keepsakes. Most of all, two crumpled bits of paper.

One, scribbled in haste, says this: 'Thank you. But no one can meet you. The army is in the street. People are afraid of them.' I never met the man who wrote this note; he pressed it into my colleague's hand when he saw us, rare foreign journalists in his country, outside his mosque on Friday. But he also saw the soldiers, their guns at the ready, and the plain-clothes spies. He was brave, but just brave enough to let us know he was scared.

My second bit of paper was discreetly passed to me by someone who works for the government – it's a message written fast and furtively by someone who wanted me to know what he was thinking, but couldn't say it out loud. 'Lost,' he wrote, 'between my morals and what they "say".' He too was brave, and worried.

After months of waiting to get to Syria to talk to people on all sides of an increasingly violent divide, I found a place where many were still afraid to speak. At times, afraid even to *say* that they were afraid. For most of this year, I have been reporting from places where Arabs have boldly declared, 'We have lost our fear.' Men and women marched down boulevards in Tunisia to stare riot police in the face. Egyptians ran into clouds of tear gas, grabbing canisters from the tarmac to throw them back at the troops. But this is a country where the army is firing on the people.

Although you don't see that in the heart of Damascus. On our first day in the maze of cobbled streets in the Old City, a shopkeeper stopped us to tell us we were lying about his country. 'Twenty-three million people love President Assad!' he shouted angrily. When I pointed out that some Syrians were protesting, he gave a precise reply: 'Well, maybe ten thousand don't like him.' A week later, I ran into him again. He was smiling this time. 'Did you tell us what you really think?' I asked 'Yes!' he insisted convincingly. Then, he suddenly remembered his sense of hospitality and scooped us a free ladleful of salted pistachios from a burlap sack. This is also Syria. That hasn't changed.

But Syria is a different country now – even if the government makes it difficult for us to see it. We didn't get permission to visit towns like Homs and Hama, where the protests have been greatest. When we were allowed to go to Deraa, where this wave of unrest began in March, we found its central square cleared of people for our visit.

This is a government that didn't want us to see much. What do they fear? In more than one place, we were warned that terrorists could attack us. Within the regime, some do believe they are fighting armed gangs; others do believe this will soon be over, and all it needs is a bit more violence to force people off the streets. But it won't be over, not any time soon.

On a visit to Douma, in the suburbs of Damascus, we were, as often happens, surrounded by a crowd, including men in shellsuits talking into telephones, watching and listening. We expected to hear the same whispered denunciations of the president, the same loud expressions of support for him. But when one student told me, 'The situation is good,' another interrupted without hesitation: 'No, it's not good, it's bad! The government is shooting the people!'

And then, an old man with a tired face made his way through the crowd to announce: 'My son was detained. His mother is crying, looking for him.' He didn't hide his face, or his anger. I asked him why he decided to tell us his story. 'I am afraid now,' he said, 'but what will happen will happen.' Then he walked away. Not long after, troops flooded into the area.

The last person we interviewed in Syria was a Christian priest. We wanted to hear more about reports that Syria's many minorities, including Christians, were worried about the future. I asked him what people in his congregation were telling him. 'It's not just us,' he said. 'Twenty-three million Syrians are afraid.'

And then our visas ran out. We stayed until the last possible moment, arriving at the border at two minutes to midnight. So the very last Syrians I spoke to were the men from intelligence. They were smiling and friendly, as most Syrians are. They just wanted to confirm that all our equipment, everything we brought in, was going out. Instinctively, I checked my pockets, ensuring those two pieces of paper were still there, staying with me.

Ian Pannell – Waiting for Godot – 28/04/12

As the state brutalised peaceful protesters, the resistance turned to violence too. The Free Syrian Army was set up in in the summer of 2011 by dissident Syrian army officers who called it 'the armed wing of the Syrian people's resistance' and openly stated they wanted to bring about the fall of the Assad regime. Many of its members were defectors from the Syrian military, sickened by orders they'd been given to shoot or bomb unarmed demonstrators. These men knew the country's power structures from the inside and were determined to do all they could to undermine them. The FSA's resolve was stiffened by the national army's use of ever-heavier weaponry, including rockets fired from attack helicopters and fixed-wing aircraft. There was fighting around the major cities of Damascus and Aleppo. By the summer of 2012, both the UN and the International Committee of the Red Cross had officially declared that Syria was embroiled in civil war. Ian Pannell met one rebel whose choice to take up arms mirrored the escalation all around him.

Abu Mohammed says he wept when he saw what the soldiers had done. He'd had to wait for two days for the troops to pull back before he dared to venture home. That's when he saw the still-smouldering remnants of what had been his life.

His flat had been set on fire. Every wall, floor and ceiling was blackened and cracked. All his possessions had been reduced to ash.

I first met Abu Mohammed in February. He lives in Idlib Province in northern Syria. I won't say exactly where, to protect his identity; Abu Mohammed isn't his real name. He's an English Literature graduate who's an unlikely recruit to the rebel Free Syrian Army: a young, thoughtful man who's more at ease discussing William

Shakespeare and Samuel Beckett than military tactics. The last time I met him he told me, 'I want people in Syria to carry a pen, not a sword.'

But conflict hardens the soul and forces people to make choices based on necessity, not want. Since I last visited, the full force of President Assad's military has been unleashed on this part of Syria. Towns and villages have been shelled, hundreds have been killed, people have been detained and homes have been burned.

'I had five hundred books – English novels and plays, American poetry – and all of them are gone,' Abu Mohammed said.

He gave me a guided tour of the charred shell of his home. I could just about make out a series of black rectangular marks on the floor of what had been his library. 'If they burned all the rest it's OK,' he said, 'but not my books. I cried when I saw my books.'

He has lost faith that the international community will help. 'Can Kofi Annan come and see my library?' he shouted, adding that the former UN Secretary General's six-point peace plan lay in the ashes of his home. He's angry and he's not alone. Those we met feel they've been abandoned by the outside world. Many asked why the West refuses to help, or how Syria is any different from Libya, where the international community did intervene.

I tried to explain that the Europeans and Americans felt they *were* helping, but that Syria was a far more complex country in the heart of a very difficult region. But it's hard to make a convincing case for nuanced foreign policy to people who've been on the losing side of a life-and-death struggle for more than a year.

In fact, for many people in Idlib, their fight with the Assad government isn't just over a year old – it's more than three decades old. A prior protest movement erupted here in the early 1980s. It had an Islamist bent and called for an end to what the Sunni majority here still see as routine discrimination at the hands of the ruling family, who're from the minority Alawite sect of Shia Islam. It was eventually crushed in a ferocious government offensive that left perhaps tens of thousands dead and many imprisoned. What is happening in Syria isn't just another offshoot of the Arab Spring; it's a generations-old movement for change. This also helps explain why the opposition hasn't yielded to the pressure it's been put under.

In a grubby tent offering little shelter from the elements, we met a group of fighters squatting around a solitary pear-shaped paraffin lamp. They were dirty, tired and beaten. The government's offensive has had an effect and, for now at least, it seems the armed insurgency is over. But the popular uprising is far from over.

Abu Mohammed and I talked a bit about the authors whose works had filled the shelves of his library. He moved from William Shakespeare to Jane Austin to Samuel Beckett – declaring that Beckett's *Waiting for Godot* was his favourite play. He wanted my interpretation: what did I think the enigmatic absurdist drama really meant? It is about two men waiting for someone who never arrive; feeling rather gloomy about what I had witnessed over the last few months, I waffled a bit about inaction and fate.

'What does it mean to you?' I asked.

Abu Mohammed smiled. 'Hope,' he said. 'I believe Godot is hope.'

Death and destruction had been wrought upon his home town and we were standing in the ruins of what had been his beloved library, yet his faith was undiminished. 'You know, we are waiting for Godot right here,' he laughed.

The people I have been talking to here share an absolute conviction that this time will be different, that what their fathers began will be finished by today's generation. And after decades of what they see as state-sanctioned oppression, the spirit of rebellion burns bright as they wait for their Godot.

Ruth Sherlock – Militants with Bread and Bullets – 14/02/13

Well before the civil war broke out, the Assad dynasty had repeatedly made the argument to its critics: it's us or the Islamists. At the beginning of the conflict, more secular, nationally focused revolutionary groups like the Free Syrian Army had been at the forefront of resistance, but as the war escalated it was more and more evident that there were religiously motivated factions jockeying for position too. The group known as Jabhat al-Nusra or Nusra Front was noticeably effective, perhaps because it was drawing on expertise from its connections with al-Qaeda and the nascent 'Islamic State of Iraq' grouping led by Abu Musab al-Zarqawi. Ruth Sherlock asked what exactly its plans – and its methods of winning hearts and minds – would mean for the Syrian revolution.

The common etiquette when conducting an interview is to sit facing your interviewee, listen carefully, and sometimes give an encouraging smile. This won't work when meeting with Hajji Rasoul, an emir or leader in Jabhat al-Nusra, a jihadist group that the United States has blacklisted as a terrorist organisation.

Wearing a beige robe, his face framed by a prayer cap and long beard, Hajji Rasoul faced forwards in the front seat of the people carrier. His security detail (seven armed men in black uniforms) surrounded the car. I sat three rows back, squashed into a small seat in the boot that put as much distance as possible between us. Respectful of his strict interpretation of Islam, I had dressed in a demure black hijab and loose clothing. But Hajji Rasoul still moved the rear-view mirror so he couldn't accidentally glimpse my reflection.

All this was in line with his strict Wahhabi views. But it was at odds with the gentle Sufi culture of Muslims in Aleppo. Six months ago when the rebels stormed Aleppo, encounters like this would have been rare. But the Nusra Front – and its hard-line beliefs – are making more and more of an imprint on Syria's biggest city.

For many in Syria the Nusra Front is synonymous with al-Qaeda. In recent months it has claimed responsibility for a litany of car and suicide bombings aimed at government targets, which have also killed hundreds of civilians. The group aspires to turn Syria into an uncompromising Islamic state governed by sharia law.

The war has turned this mercantile city, once a trading hub for gold, jewellery and expensive silks, into a place of depravity and despair. Mounds of rubbish lie rotting on the streets; families live in bombed-out ruins; the wounded are left to beg for their next meal.

Furious at what the main rebel group, the Free Syrian Army, has done to their livelihoods, Aleppo's destitute people are turning to the Nusra Front for help.

Already known for having some of the bravest fighters, in the last few weeks the group has started humanitarian programmes. When grain stores first fell into rebel hands the supply of bread to Aleppo all but ceased. Now Jabhat have monopolised control of the warehouses and are carefully managing the supply. 'We have four factories working and each has the capacity to make 150 tons of flour per day,' said Hajji Rasoul. 'We are able to feed all of the liberated territories in Syria.'

The operation is impressive. In one Aleppo district I watched as bakers tirelessly churned out Jabhat's piping-hot flat bread and sold it at a bargain price. In this area alone more than 26,500 families are grateful recipients of this aid. The Nusra Front's other projects include subsidising farmers to replant wheat fields; bringing oil and gas to Aleppo; restarting medicine factories and organising rubbish collection.

I asked the bakery's manager why he thought Jabhat al-Nusra had succeeded in these projects when other rebel groups are still barely able to sustain themselves, and what he thought their motivation was. Smiling, he pointed to the sky and replied with one word: 'Allah'. God.

The projects are quickly winning hearts and minds among Aleppo's residents. The Nusra Front and their fellow Islamists have a reputation as pious men: an attractive quality in a city where murder, looting and kidnapping for ransom are rife. 'They provide us with heating and bread, and they are good Muslims,' said one father of nine. 'Eighty per cent of the FSA are rotten.'

Hajji Rasoul was coy about the group's vision for a future Syria. But in Aleppo's rebel-held areas, evidence of an Islamist social template is emerging. Most of the humanitarian projects are coordinated through a strict sharia court that Jabhat al-Nusra controls in partnership with three other hard-line groups.

The court regularly arrests other rebel commanders. Some are accused of looting or executing their prisoners, but 'drinking alcohol' or having 'relationships with women' are also deemed worthy of punishment.

Sitting in a safe house close to the front line, FSA commander Abu Hossam painfully lifted his trousers to show me deep black bruises across both of his lower legs. He was imprisoned and tortured in the sharia court's prisons for thirty-six days before he received the verdict of 'not guilty'. 'Smoking marijuana, spending time with women and starting a nightclub were some of the charges they brought against me,' he said.

Islamist prowess is a source of seething resentment for many FSA groups and secular Syrian citizens alike. During popular protests, calls for support of pro-democracy rebels still outnumber cries for the 'Islamic army'. But whilst the Islamists are better funded, better organised and better armed, there is little the FSA can do.

Jeremy Bowen – What We Can and Can't Cover – 27/05/13

The ever-present ethical dilemmas of war reporting sharpened even further for those trying to cover the Syrian conflict. When, if ever, is it justified to pay for a source, or intrude on someone who's injured or grieving? How graphic can or should journalists be about the horrors they see? And which risks is it worth running for a reporter or a news organisation? Jeremy Bowen considered what it took when trying to get anywhere near the truth in Syria.

When I started learning how to be a reporter, one of the highest forms of praise was to be talked about as a 'good operator'. My colleagues said about one man, who was a well-known television reporter twenty-five or thirty years ago, that he couldn't write his way out of a paper bag, but he was a great operator. He won prizes and glory, not for his scripts, which were short of nuance or long on clichés, but because they were good *enough* – taken together with the fact that he was always in the right place at the right time to do a supercharged piece to camera, and for his crew to get dramatic pictures.

In Syria, as in every war, reporters, producers and camera crews need to be good operators. They need to find ways to get to difficult and dangerous places. In Syria, as in every war, that can mean compromises, and sometimes deals with people you might otherwise prefer to avoid.

The challenge in reporting wars is to survive the day at the same time as coming up with some decent journalism. There are often surreal moments. In Damascus we stay at an extremely comfortable hotel which has become the base of choice for the UN as well as journalists. Before one trip to the rebel-held part of the city I prepared with a visit to the gym and a massage. A couple of hours later I was in the bombed-out remains of a suburb, watching a couple of dentists trying to treat a man whose feet had been blown off by one of the regime's rockets. Everyone in the improvised dressing station was nervous about the odds of more rockets falling on us.

I have been concentrating, visas permitting, on Damascus. Getting over the border from Lebanon, once the visa has been stamped into your passport at the Syrian embassy in Beirut, means transporting one of the Assad regime's secret policemen down from Damascus to search our gear. I wouldn't say we're old friends yet, but it helps that the same man usually does the job, so it's easier to be friendly. When the search is finished and we drive the thirty miles or so to Damascus it is good to have the secret policeman along because it means the BBC's minibus gets waved past the men with guns at roadblocks.

Is this a way of conniving with the regime? I don't think so. Journalists, in peace and war, always have to negotiate to get interviews. In Damascus we have a minder some of the time who reports back to the ministry of information. But he also gets us to places that we couldn't get to without him – like military hospitals to see Syrian army casualties. When we need to see the rebels in the suburbs, we use our own connections. Dealing with them also requires negotiations and compromises. Sometimes, on both sides, there are sights they don't want us to film.

Moving around war zones means getting permission from the people who control the territory you're in, whether it's the Assad regime, or the Nusra Front, who are the most efficient rebel fighting force. It makes no difference that President Assad runs one of the world's most condemned regimes, or that the Nusra Front supports al-Qaeda and is listed as a terror group by the United States. The deals, and permits, are stepping stones to get to the story.

It's getting harder, though, because the war is tearing Syria into pieces. More chaos means more difficulty and danger for journalists. And because these days, winning the media battle is so important that journalists can become targets – not because they're in the wrong place at the wrong time, but because of who they are and what they do. Foreigners stick out, and make valuable hostages.

The troubles faced by news teams, though, are as nothing compared to the torment the Syrian people experience. A foreign passport can be a liability, but it can also be a protection. We are not having our families killed and our streets and towns burnt or pounded into rubble. Best of all, we have homes and lives in safer places. Syrians do not.

Iraq

Jim Muir – Papa's with Jesus – 11/11/10

*The steady dwindling of Christian communities across the Middle
East is a long-term trend. Before the First World War around 20 per
cent of the region's population belonged to one of the dozens of long-
established Christian sects which took root there. Every country, apart
from the Gulf states, had a significant Arab Christian minority, living
in various degrees of comfort with majority-Muslim rule. Today, that
proportion is down to around 4 per cent. Emigration, official disinterest
and harassment have all played their part. But there were also surges
of concentrated persecution, and Christians in Iraq felt particularly
vulnerable. Before the US intervention and the fall of Saddam Hussein
in 2003, there had been around 1.5 million Christians in Iraq, but relentless sectarian attacks and the fear of Sunni fundamentalist groups
have since reduced them to under a quarter of a million. Jim Muir
saw a community shaken to its core by a terrorist assault on Baghdad
Cathedral in which fifty-eight worshippers, priests, policemen and
bystanders had been killed by jihadist suicide bombers.*

We were in one of the churches near where we live in central Baghdad, and they were preparing to bury some of the dead from the
disaster in the cathedral just a stone's throw away. There was already
a row of coffins, draped in Iraqi flags, lined up in front of the altar.
Every so often, the doors would open, and more coffins would be
carried in one after the other, bringing fresh waves of sobs and
applause from the grieving congregation.

    On her knees, slumped over the end of the coffin laid out closest
to me, was a young mother who had just lost her husband. Their
bright little daughter, less than two years old, was playing on the

coffin, and running back and forth – obviously just too young to understand. Every so often she would ask her mother, 'Where's Papa?' And the reply would come: 'Papa's with Jesus.' You could tell from the dull, faraway look in the mother's eyes that she was looking back on the life she was living, that had just been abruptly ended, and ahead to a new, harder life, as a widow with children to bring up, in a place where life and loved ones could be taken away at any moment without warning.

Perhaps it was a small comfort to the mourners that among the many who came to honour their dead were people and leaders from other communities, Sunni and Shia Muslims, politicians and clerics. What happened at the cathedral sent a huge shock wave through their community, but it was also felt by all the others, despite the many unspeakable atrocities they themselves have suffered in recent years. Think of the many Shia mosques and shrines that have been bombed, the funeral processions and pilgrim groups attacked, the markets and crowded streets devastated, or the thousands of Sunnis abducted, tortured and killed just for being Sunnis.

But many instinctively felt that the attack on the church was something special. The Christians are a minority, less than 5 per cent of the population. They're not involved in the struggle for mastery of Iraq. So for most Iraqis, attacking them is, well, *haram* – which means not just forbidden, but also something that shouldn't happen – something reprehensible and pitiful and wrong. Because the Christians are a threatened and dwindling minority, many probably also felt that this was an attack on the very fabric of Iraqi society, on the coexistence that has to survive if Iraq itself is to survive.

It was Bishop Metti Metok's cathedral that was attacked. Two of his priests were killed, and half his regular congregation left dead or wounded. When his flock ask him now if they should stay or try to leave, he says they should stay, to bear witness to their faith. But he says, sadly, he knows he can't stop them going. Everybody knows that nobody is truly safe. The government may step up security around churches. But the community is scattered everywhere, impossible to protect.

Many of them are already going. Fadya and her family are leaving this week, for California. They've been waiting two years for their American visas, so it's not a spur-of-the-moment thing. She says she just wants not to be afraid any more – afraid for herself and for her four- and eight-year-old boys. As well as Arabic, the family speak their own Assyrian dialect, as the community here has since biblical times. But they have hardly a word of English between them. These are not posh people, not middle-class. Fadya worked for us as a cleaner. She's never been on a plane.

It's sad to think that if I ever see Fadya again, she'll be different, and her boys Issa and Edmond will be little Americans. Her mother's still here, and her brothers. But other Iraqi Christians are leaving every day. Perhaps out of the current intense political struggle will emerge the foundations of a stable future. That's all that the Christians who remain behind can pray for, so that these ancient communities can cling on, and eventually, people like Fadya can think about coming back.

Gabriel Gatehouse – A Responsible Drawdown – 04/09/10

During the intervention of 2003 and the years of occupation and interim administration which followed, 4,497 members of the US military died in Iraq. The war turned out to be a disaster for overconfident would-be nation-builders, and attracted bitter criticism from allies in the West, as well as across the Muslim world. The American public grew disenchanted with their country's presence in Iraq, asking pointedly what all the billions of dollars spent and all the goodwill lost had gained for the USA. Voters still wanted to support their troops – but most of all, to bring them home. How to square that circle, and look tough while still leaving, was a problem for both foreign-policy wonks and politicians. As he covered the withdrawal of the last US combat troops, Gabriel Gatehouse wondered how much euphemism was at work.

As the late-August sun rose over an American military base on the outskirts of Kirkuk, I sat on the step of a Portakabin sipping my morning coffee. The mug was emblazoned with the words: 'Operation New Dawn'. A day before the official end of combat operations, the commemorative merchandise was already on sale.

I'd stayed on this base before, in October last year. Then, it was teeming with soldiers. Now, the place felt deserted, the canteen half empty, only the odd armoured vehicle rumbling its way along the dusty lanes of the camp. Since then thousands of US troops have left this base, sent back home as part of America's so-called 'responsible drawdown'.

In the lexicon of the US military, a 'drawdown' is the opposite of a 'surge'. Removing tens of thousands of soldiers from Iraq is not a withdrawal – that would imply a lack of commitment, a loss of interest. It's most certainly not a 'retreat', for that would imply defeat. No, it's a 'drawdown', and a 'responsible' one at that. In official speeches and press releases the two words always go together. And how fitting that the final chapter of America's military involvement in Iraq should have its own stock epithet, just like the characters in the great Homeric epic myths: 'swift-footed Achilles'; 'wily Odysseus'; 'Hector of the Shining Helmet'.

A day later, on the outskirts of Baghdad, on another American base known as Camp Victory, the assembled foreign press was herded into one of Saddam Hussein's old palaces. In a marble-clad central hall with a massive chandelier, soldiers and dignitaries watched as a flag was solemnly passed from one American general to another. And with that, seven years of combat in Iraq, a mission known as 'Operation Iraqi Freedom', segued seamlessly into 'Operation New Dawn'.

That ceremony was Obama's 'mission accomplished' moment. President Obama didn't appear wearing a flight suit, as George Bush did in 2003, to proclaim victory on the deck of an aircraft carrier in the Gulf. But the careful choice of words filters down to the lower ranks, and to the press, and through them to the American people. The message was: job done.

Before the ceremony began, around a dozen officers were on hand to rotate around the hall giving ten-minute interviews to the media – a sort of speed-dating with generals. I asked one of them whether it had all been worth it. He said he thought it had. He talked about 'the overthrow of a dictator and all that he stood for, all that he did to hurt the people in Iraq. You have now had two elections,' he said, 'you have a democratically elected leadership . . .'

A few days earlier, I had put the same question to an Iraqi in his late twenties. After the invasion he had worked for the Americans as an interpreter, until death threats forced him to quit. But he is still an Americophile, a heavy metal fan who plans to move to Texas and start a new life. Why was he leaving, I asked? Because, he said, he didn't feel free in Iraq. 'Under Saddam Hussein we were more free,' he said. Free to wear the clothes he liked, free to listen to the kind of music he wanted. Nowadays, he said, those kinds of choices could get you killed.

The idea of freedom, with a capital F, is very important to the Americans here. The military bases are festooned with banners, sent to Iraq by grateful schoolchildren back home, thanking the troops for the 'gift of freedom'. It is this gift that, many believe, they have also bestowed on the Iraqi people. The toppling of Saddam and the holding of democratic elections attest to the fact.

The transition from 'Operation Iraqi Freedom' to 'Operation New Dawn' implies that goal has been achieved. Yet the reality is this: there are still nearly 50,000 American troops in Iraq. And Iraqis are still being killed in large numbers. Iraq was not Obama's war. He promised to end it and to bring the troops home. But in order to do so with America's pride intact (and with an eye on the mid-term elections) the USA is building a myth around this departure, using words rather than guns.

This process has been under way now for some time. A few months ago, an aide to a high-ranking American general invited me on something called a 'peace-field rotation'. When I asked what that was, the aide responded that they were trying to phase out the word 'battlefield' in Iraq now that security has improved.

Not everyone is convinced. On our way to the ceremony, we had our documents checked by an American soldier. 'You guys make sure you do a good job now,' he shouted after us, grinning in the scorching sun of the Iraqi summer, 'of reporting the absolute zero change that's going on here today!'

Aleem Maqbool – Karbala: Shrines and Battles– 25/05/13

As one of the holiest sites in Shia Islam, the Iraqi city of Karbala has been repeatedly targeted by extremist Sunni groups which abominate the sect. Even during Saddam Hussein's era, it had been a religious counterweight to the secular Ba'ath Party power structure in Baghdad, and thus a hotbed of Islamist resistance. After the Gulf War in 1991, Shias in Karbala and Najaf had risen up against Saddam in an abortive uprising. After the US invasion of 2003 and Saddam's death, the traditional Shia pilgrimage to Karbala was bombed by Sunni jihadists in 2004, and it would be attacked again. Aleem Maqbool journeyed to the heart of a sectarian division which would only become more intense in the years to follow.

It is a privilege to do this work. I kept thinking that as we travelled along the road from Baghdad, to the city of Karbala.

Just like Bethlehem and Nazareth for Christians, Karbala is one of those places Muslim children hear about from when they're very young. For many it takes on a mythical, unreachable status. But here we were pulling up to the main security checkpoint of Karbala, being asked to park on one side while our papers were checked. Many pilgrims were going through on foot.

As we carried on, the gold dome of the Imam Hussein mosque rose from the centre of the city. Soon we found ourselves standing at one of its many ornate doorways. I watched a little girl pull back her mother by the hand and chastise her for not having kissed the entrance in respect. Her mother dutifully spent a moment pressing her lips against the huge wooden doors to the mosque, before going inside.

In the vast prayer hall with its gold and marble, its huge chandeliers and its intricate blue-and-white tiling, was an incongruous red neon sign. It marks the spot where it's believed Hussein, the grandson of the Prophet Muhammed, was beheaded. The area on which the mosque complex is built is thought to be the site of the Battle of Karbala, in which not only Hussein, but seventy-two followers and members of his family, including his infant son, were slaughtered. When you see Shia Muslims, bloodied, whipping themselves in their annual processions, they are commemorating the events that took place in Karbala. Because while Muhammed's grandson is revered by all Muslim sects, Shias trace their beliefs directly to his teachings.

The Shrine of Hussein, in the centre of the mosque, is now circled by a constant stream of pilgrims, kissing the marble, praying, often shedding tears. For Shias, this shrine represents the greatest tribute to martyrdom.

That battle of Karbala, in the seventh century, in which Hussein was killed, is often cited as the moment Shia and Sunni Muslims were cleaved apart. But Friday prayers in the Imam Hussein mosque looked almost the same as prayers in a Sunni mosque. There are small differences in the rituals: at one point, for example, instead of crossing their hands over their stomachs, the lines of Shia devotees kept their hands by their sides. But there is difference enough, it seems, that even to this day, some feel the need to continue the slaughter.

Over the last weeks, Iraqis have witnessed the kind of sectarian violence they'd hoped was a thing of the past. One morning eight bombs went off in an hour in Shia districts of Baghdad. On another there were eleven, almost simultaneously – again in Shia neighbourhoods. There was an attack on a Sunni mosque in Baquba, too, and another close to a funeral procession.

Speaking to Iraqis at the bombsites and in hospitals what surprised me was the very apparent *lack* of hatred towards the other sect. Instead, there appeared to be resigned consensus that Iraq's politicians were exploiting sectarian differences for their own gains – but that so too were foreign powers.

It's something that has created a huge Sunni–Shia rift across the region – Iran the big power on one side, Saudi Arabia and Qatar on the other. Other countries in the Middle East, with mixed populations, are feeling the effects, as Sunni and Shia groups within them start to align themselves in what many now see as a holy war.

In Iraq, the links between Sunni militant groups like al-Qaeda and the rebels fighting across the border in Syria have been much talked about. But a couple of days before going to Karbala, we met a man who was involved in recruiting *Shia* fighters to go to Syria – not to fight with the opposition, but alongside President Assad's forces. 'When somebody attacks your beliefs,' he told me, 'you have to defend them.'

And where, I asked, are the Shia fighters going in Syria? His answer took us right back to Karbala. He said their main aim was to defend the shrine of Zainab – another of the Prophet Muhammed's grandchildren whose brother and two children were among those killed in the battle of Karbala. After which, Shias believe, she was taken to Damascus, where she later died. Sunnis disagree and think she was buried elsewhere, and in recent months there have been many reports of attempts to attack the Shi'ite shrine. The Shia recruiter told me they would keep laying down their lives until the shrine of Zainab was safe.

As I left the mosque in Karbala, carrying the traditional memento of a small amount of earth from its grounds, it was hard not to reflect on how events there fourteen hundred years ago are shaping the world today.

Iran, Yemen, Saudi Arabia

Rana Rahimpour – Iran – Toothpaste in Tehran – 12/05/18

The BBC has very little chance to hear direct from ordinary Iranians, as Iran's government does not allow it to operate a bureau in the country and forbids local reporters from 'stringing' (that is, reporting part-time) for foreign news organisations. So employees of BBC Persian, the Farsi-language service which broadcasts around the world on TV, online and on radio, have to work from a base in London. In some cases their relatives have been detained in attempts to exert pressure either on the BBC or on the UK. BBC News crews visiting Iran for short periods, whichever language they use and wherever their reports will appear, usually have to work under close supervision while in the country. More often they aren't allowed in at all. Still, as relations between the US and Iran went into a newly combative phase under President Trump, Rana Rahimpour offered a concrete example of what tougher American sanctions could mean on the streets of Iran.

Like most Iranians living abroad, I often send presents to my parents back home: shortbread biscuits, English breakfast tea or After Eight mints. But back in 2012, when I phoned to ask my father what he would like from London, he said: toothpaste!

'Toothpaste?' Yes, I'd heard him right. 'But why?' I asked. Because Iranian toothpaste just didn't taste right, he said, and imported toothpaste was nowhere to be found.

Toothpaste quickly became one of their most cherished presents. My parents would buy large packs of it every time they came to London, to take back as gifts for my aunts and uncles.

Two years later, I met up with some relatives in Turkey – safe territory for me, a journalist working for the BBC Persian service (and therefore seen as an enemy of the Iranian state) and a perfect holiday destination for Iranian nationals, who don't need a visa to visit.

On the first evening of the holiday, I realised I had forgotten to bring my own toothpaste so asked my sister-in-law for some of hers. It was then that I understood what my father had meant. The tube looked like any other. The paste itself was a familiar greenish-blue, though it looked a bit dry. But when I put it in my mouth, it was like eating a spoonful of flour. My teeth felt heavy, not clean – as if a layer of dust had settled on them.

Iran has been under US sanctions since 1979, after supporters of the Islamic Revolution took US diplomats hostage. When I was growing up in Iran, America was a faraway land that, we were told, was our enemy. Every weekday, in our white headscarves or *maghnae* – part of the school uniform – we had to stand in line and chant 'Death to America'.

During the eight years of war between Iran and Iraq (in which the West supported Saddam Hussein) we had to line up again. My father and I queued to buy rice and cooking oil – with coupons, because food was rationed. In the evenings, we'd gather in front of the TV to hear the names of the martyrs who'd died on the battlefield. The announcements were followed by patriotic songs, eulogising those who had been killed by Saddam and his allies. 'Death to America' was a popular slogan in those years.

For many the election of President Mohammad Khatami in 1997 felt like a breath of fresh air. He was a Western-educated reformer who dared to talk about freedom, was welcomed in European capitals and spoke about dialogue between civilisations. But during his second term in office, President George W. Bush named Iran part of his 'axis of evil'. Doors were slammed shut before they'd even been opened.

More international sanctions followed, the harshest in 2012. Iran's president at the time was Mahmoud Ahmadinejad, a hard-line populist who called them 'worthless pieces of paper'. In reality, they severely affected Iran's economy. Iran was prevented from selling its oil to most countries, and even those which were still buying it couldn't pay in cash. Patients needing imported medicine had to make do without. Inflation reached 30 per cent, and economic growth turned into economic decline. But all this did eventually bring Iran to the nuclear negotiating table.

In July 2015, after eighteen months of gruelling talks, it was finally announced that an agreement had been reached. Images of US Secretary of State John Kerry shaking hands with his Iranian counterpart, Javad Zarif, were watched with disbelief by many in Iran. Upon his return to Tehran, Mr Zarif was welcomed at the airport as a hero and a messenger of hope.

My father stopped asking me to send toothpaste. He said he could find it in Iran, for the same price as in London. For a while, life seemed to be back to normal.

But hopes were quickly crushed. After the election of Donald Trump, doubts about the future of the deal grew. Iran's economy, which had barely recovered from the sanctions, was hit by a fresh wave of uncertainty. Even before Trump's decision to withdraw from the agreement, the Iranian riyal had lost half its value. On Tuesday, the familiar chants of 'Death to America' could once again be heard from hard-line MPs in parliament and activists in Tehran's Azadi Square, taking us back to square one.

When I phoned my father the next morning, he jokingly said, 'It looks like we are going to need you to send toothpaste again.'

Frank Gardner – Yemen – A Many-Layered Conflict – 04/04/15

Ali Abdullah Saleh, who'd served as president of Yemen since 1990, had kept on top of its intricate regional politics with a combination of guile and force. He'd always managed to find leverage over secessionist movements, ambitious rivals or recalcitrant clans. He once described governing Yemen as 'like dancing on the heads of snakes' – but even his fancy footwork couldn't keep him in power for ever. As mass protests in spring 2011 forced him to make concessions, he bargained, prevaricated and postponed every step of the way – until a bomb attack on his compound in June left him seriously wounded. Even then, he didn't formally cede power to his successor, Abd Rabbuh Mansur Hadi, until early the next year. And despite his injuries, Frank Gardner learned that Saleh was still plotting and pulling strings years later, as Yemen's national crisis worsened.

The Romans had a name for Yemen. They called it Arabia Felix – Happy Arabia – because of its lush, rain-fed mountain scenery. Today that epithet sounds tragically inappropriate. Already the poorest country in the Middle East – racked by soaring unemployment, dwindling oil and water reserves, and home to the most dangerous branch of al-Qaeda – Yemen is now being torn apart by a war of many sides.

The Saudi-led air strikes began last month, raining down precision-guided missiles on a rebel group called the Houthis, who swept down from their mountain stronghold in the far north six months ago, taking town after town, and pushing out the UN-recognised President Hadi. That alarmed the Saudis and the other Gulf Arab states, especially as they suspect the hand of Iran is behind the Houthis' spectacular blitzkrieg. How else, Saudis keep saying to me, could an impoverished group of tribesmen get the training, the weapons, and the money to take over half the country?

There's a sectarian angle here too. The Houthi rebels are Zaidi Shias, like around a third of Yemenis. The Saudi rulers are suspicious of Shias, many of whom look to Iran for spiritual leadership. Saudi Arabia is predominantly Sunni and the Saudis are starting to think they're getting encircled by proxies of Iran wherever they look: in Lebanon, Syria, Iraq and now in Yemen. Enough, they said, drawing a line in the sand. At a secret summit in a Saudi palace last month, they threw together a ten-nation coalition in a belated and possibly doomed Gulf Arab attempt to turn back the Houthi takeover of Yemen and restore their ally to power.

But in fact the Houthis largely owe their military success to someone much closer to home. They've formed an alliance of convenience, a sort of pact with the Devil, with the very man who tried to bomb them out of existence five years ago. Ali Abdullah Saleh ruled first North Yemen, then a unified Yemen, for thirty-five years, until he was forced out of power by the Arab Spring protests. He refused to believe that Yemen was better off without him. So he set about wrecking the peaceful political transition of power that the country's friends had worked so hard to engineer. Whole units of the Republican Guard remained loyal to him, bombs went off and towns were fought over. President Hadi, the elderly, genial southerner who replaced him, has been no match for Saleh's machinations. He must be rueing the day he agreed to let his predecessor stay on in Yemen.

I interviewed Saleh once, in his fortified palace in the capital, Sana'a. It did not go well. Speaking in Arabic without a translator, I asked him what he wanted his legacy to be. The unification of North and South Yemen, of course, he replied – the crowning achievement. I thought I would soften him up by asking what benefits this had brought, but the way I said it in Arabic came out as, 'Well, what was the point of that?' 'WHAT?!?!' he barked, glaring at me furiously, summoning his official translator and looking pointedly at his watch.

As president, Saleh fought six short wars against the Houthis, which ended in an uneasy truce. Now he's cynically using them to destroy those who he sees as usurping his power. The Houthis are fierce, effective fighters, used to living on very little in the black, volcanic mountains that straddle the Saudi–Yemeni border. When I visited them in their stronghold city of Sa'ada, I was woken on day one by a burst of heavy machine-gun fire from a pickup truck outside in the street. 'It is celebration,' said the man on reception, unfazed.

Later, I met a pair of brothers who took me out to the mountains to show off their skills with a Kalashnikov. Chewing the narcotic qat leaf and racing across the desert in a beaten-up old car, they thought it was hilarious to swap places behind the wheel while doing sixty miles an hour. Their shooting was every bit as wild as their driving and before long a farmer emerged, shouting and cursing. 'What the hell are you doing?' he raged. 'Bullets are coming down all around my sheep!'

I have no idea what dizzy heights those two rose to after that in Yemen's tribal hierarchy; but the Houthis and their allies are now in control of most of the important parts of Yemen, despite more than a week of airstrikes. If those fail to dislodge them then the Saudis have not ruled out a ground invasion, but everyone knows that carries enormous risks of getting bogged down.

The Houthis also face a more dangerous foe: the jihadists of al-Qaeda. They are Sunni fanatics and hate all Shias, including the Houthis. In Yemen, al-Qaeda seems to be the only force capable of confronting the rebels on the ground. Recently its ranks were swollen by a jailbreak of dozens of convicted al-Qaeda fighters. Soon they will be rallying the Sunni tribes to join forces and fight the Houthis from the north.

All the while, the Americans look on from afar, in despair. Until just a few weeks ago they thought they had a reliable partner in President Hadi. President Obama even held up Yemen as a shining example of a counter-terrorism partnership. Now that partnership has crumbled to dust, and so too have Yemen's immediate hopes of emerging from this intractable nightmare.

Bethan McKernan – Yemen – Profits of War in Marib – 09/12/17

Nearly a hundred thousand people have been killed by Yemen's civil war since it began in March 2015. Tens of thousands have been injured, millions displaced and millions more are on the verge of starvation. At the end of 2017, the Houthi rebels backed by Iran were continuing their fight with regional forces allied to President Hadi. While Hadi was in exile, his Saudi backers bombarded Yemen at the head of an international coalition (including Egypt, Bahrain, Sudan, Kuwait and Qatar, with considerable financial and air support from the UK and US) trying to restore his government. Ordinary Yemenis found themselves trapped in the middle of a proxy war in which some of the region's deepest enmities were being played out. Even so, Bethan McKernan found some locals were prospering amid the wreckage in a city still loyal to the president.

Before I left for Yemen, my mother asked whether the government or the rebels were the 'good guys' in the war. I had to pause before answering. 'It depends which way you look at it,' I told her. I ended up having to repeat this to myself throughout the trip.

Yes, Yemen has been ravaged by two years of civil war. The grisly death of former president Ali Abdullah Saleh really has opened a new chapter in the conflict. The only certainty is that most Yemenis will continue to suffer. But in Marib, the province I visited, the opposite is true. It's still loyal to the government of Abd Rabbuh Mansur Hadi – even though he's fled to Saudi Arabia. The rest of the country has descended into chaos, but Marib city is actually thriving.

Breezeblock homes are rising out of the sand at a ferocious pace. Astroturf from Germany will cover a new football pitch. Exotic foods like hamburgers are available in restaurants; there's even a Baskin Robbins ice cream parlour. Google Maps has already (rather optimistically) labelled a strip of wasteland as the new airport. But for Marib's ambitious governor Sultan al Ardah, the sky is the limit.

The town can thank him as well as the province's oil reserves for its good fortune. Sultan is the kind of man who still seems tall even when he's sitting down. He spoke fondly of visits to New York and Paris, but was not so keen on London. Once, he complained, he'd been turned away from dinner at the Ritz for wearing a *thobe* – the Arabian robe – rather than a shirt under his dinner jacket.

It was Sultan who'd arranged for a group of Western journalists to make a rare visit to Yemen, secured them visas, and hosted them while they were in the country. The governor was keen to show off his wealth and power, laying on several feasts of camel and lamb – which made me feel very guilty, knowing more than seven million people in Yemen are facing famine.

With a brother on the USA's terror list for allegedly funding al-Qaeda, Sultan's role in the conflict is far from black and white.

He categorically denies the claims against his brother. For the time being, though, in Marib his word is law. One afternoon, as I was sitting in Sultan's residence while he presided over a khat chew, his mobile rang. From what I could hear of the conversation, it appeared two local tribes were at loggerheads over a murder and needed him to arbitrate. Sultan listened for a minute or so, pushed his glasses up his nose then ordered one tribe to give the other four AK-47s as a way of making amends. Justice was served in two minutes flat. One particular brand of it, anyway.

Marib may be booming but the area is not free of violence. The war was always at the periphery of my vision. I glimpsed destitute refugees camped out on building sites. A rocket landed on the outskirts of town during my short visit; we were told the Houthi rebels had fired it. And on the way out, the Saudi-led coalition – also raining missiles onto Yemen – shut down all the borders, which meant driving a very long way through territory controlled by al-Qaeda to the border with Oman.

Amid the tragedy and bloodshed, though, there are opportunities for the savvy tribal leaders, arms traders and khat dealers who know where to look. Chewing the mildly narcotic khat leaves is Yemen's beloved national pastime. It has also been blamed for worsening food security. Farmers can make more money from khat than wheat or vegetables. Even before the war broke out, 90 per cent of Yemen's food had to be imported.

Now, despite the conflict, truckloads of khat don't have any problem getting from farms on one side of the front line to markets and customers on the other. Vehicles carrying anything else – guns, aid, people – are regularly held up, but 'No one is going to shoot the khat driver,' said dealer Bashir Samedi. He was wearing a spotless white *thobe*, a traditional curved *jambiya* dagger and a constant grin. 'People in the khat business have a kind of diplomatic immunity nobody else in Yemen enjoys,' Bashir told me. As he smiled, bits of green khat oozed through his teeth. 'The UN could learn a thing or two from us if they want to get aid places,' he added.

After all, he's now one of the wealthiest men in town – and what passes for good, bad or even normal in this war just isn't very clear any more.

Sebastian Usher – Saudi Arabia – Brazening It Out – 27/10/18

Saudi involvement in the conflict in Yemen was just one of its actions to provoke international condemnation. The dynamic crown prince, Muhammad bin Salman, also became the Saudi defence minister in 2015, taking on a crucial role in the war in Yemen, as well as in setting the course for Saudi Arabia's future. His public image as a reformer was carefully cultivated, with a flurry of initiatives like allowing women to drive, reining in the religious police and allowing public entertainment. But dissenters could still expect to suffer. Western diplomatic allies were often reluctant to criticise Saudi Arabia on matters of human rights at home – like public executions, abuse of migrant workers or the enduring legal system of male 'guardianship' of women. There was simply too much trade and strategic influence at stake to risk good relations. But assassinations abroad, like the murder of journalist Jamal Khashoggi, apparently inside the Saudi Embassy in Istanbul on 2 October 2019, were harder for the kingdom to spin. Attending the so-called 'Davos in the Desert' investors' summit soon afterwards, Sebastian Usher was struck by the unapologetic tone.

The vast facade of the Ritz Carlton rises from the escarpment on the road to Mecca. But the lavish conference just staged here was meant to be all about the future of Saudi Arabia, not its past. In the end it was neither; instead it was deadlocked in the very tricky present which the country has been experiencing since the horrific killing of Jamal Khashoggi.

All the focus from the very start was on one man, Crown Prince Mohammed bin Salman. Rumours abounded about when he might appear. Then on the afternoon of the first day, there was a hubbub in the hitherto subdued conference centre – overshadowed not just by the global outcry but by the looming weight of gargantuan chandeliers in the finest desert-Versailles style.

It was just a flying visit by the crown prince, enough time to pose for selfies and take a seat beside the king of Jordan and then he was gone. But the aim was clear – to show the world that Mohammed bin Salman was not hiding away.

Despite all the cancellations by big-name speakers and sponsors, the conference was still far from deserted. Thousands thronged the ornate corridors in search of coffee, cake and deals. There were plenty of Saudis about. The mood, although subdued, was pragmatic and resolute, the horror of the reported details of Mr Khashoggi's murder a regrettable failing on the part of the hosts, but not – in the end – a deal-breaker.

It increasingly resembled an expensive bubble of unreality, at odds with the sharply rising chorus of revulsion outside. This was only heightened by what seemed a seriously misjudged and heartless attempt at PR by the Saudis, with a gruesome photo opportunity where the crown prince shook the hand of Mr Khashoggi's son – who was not himself at liberty to travel at the time, but has since been allowed to leave the country.

At the gala dinner on the first night, I found myself wandering around the indoor pool surrounded by tables groaning with canapés, thinking of the very different night being experienced by some of those who have helped this country on the road to reform in recent years – only to find themselves in jail now. Eman Nafjan, Loujain al-Hathloul and Hatoon al-Fassi had all seemed to embody the new Saudi Arabia that the crown prince has said he's trying to produce. Their reward was solitary confinement. I scrubbed their names from my contacts list before arriving in Riyadh, just in case being on it might have caused them more harm.

The atmosphere on the second day was darker. One previously buoyant American investor looked haunted as he rushed to an urgent meeting to discuss his billion-dollar deal. All day, anticipation grew over how the crown prince would finally broach the Khashoggi affair. The minutes before he appeared were like waiting for a rock star. The packed auditorium cheered him as he entered.

The moderator took a deep breath and asked the unavoidable question.

Mohammed bin Salman's response was a bravura display of sorrow, defiance and suspicion. Most notable were his warm words for Turkey and its president, Recep Tayyip Erdogan. This, of course, is the man who has been expertly turning the screws on the crown prince for weeks: a strategy designed to put a rising regional rival, politically and ideologically, in his place, while at the same time extracting a high price from the Saudis. The immediate speculation in the room was that the two men might now have reached an understanding on how to prevent the crisis turning into a true tipping point for Saudi Arabia.

By the end of his appearance, Mohammed bin Salman was back to his laughing, backslapping self, projecting a strong sense both of bonhomie and power.

Sitting by his side was Saad Hariri, the Lebanese prime minister. The two have history. In an infamous broadcast last year (which his fellow Lebanese swiftly dubbed 'a hostage video') a sallow, nervous Hariri made a shock announcement of his resignation, all the way from Riyadh. The diplomatic fallout was huge and Mr Hariri eventually returned to Lebanon and rescinded his statement. The two men appearing on the same stage a year later did raise eyebrows.

But as if all shadows were now dispelled, the crown prince dismissed the incident with a quip that electrified an audience desperate to have their faith in his powers restored. Mr Hariri plans to stay a couple more days in Riyadh, the prince said, 'but don't worry, we're not kidnapping him'. The attendees jam-packed into every available space laughed and cheered.

The Saudi hope now is that the crown prince's mix of bluster, brazenness and charm will be enough to start turning the page on the killing of Jamal Khashoggi, whose terrible end was all but forgotten in the marbled halls of the Ritz.

ISIS

Caroline Wyatt – Kurdistan Is Creaking – 16/06/14

In 2014 a new extremist group burst out of the tangle of jihadi networks in Iraq to become the most extreme threat to regional security in a very crowded field. It called itself Ad-Dawlah al-Islamiya, the Islamic State, though others preferred to label it IS, ISIS, ISIL or Daesh. Its worldview was an apocalyptic Sunni vision of a Middle East cleansed of corrupt monarchies, godless one-party states and boundaries drawn by European colonialists. It also wanted to eradicate Shias and other religious minorities (who it argued were heretics) from the areas it took over, and to enforce its own extreme interpretation of sharia law on any Sunnis living in its self-declared caliphate. Its takeover of the major Iraqi cities of Fallujah and then Mosul drove tens of thousands to run for their lives. Caroline Wyatt felt a rising drumbeat of dread in Iraq's semi-autonomous Kurdish region, where many of them took refuge.

In Mosul, smokers are getting desperate. Not content with crucifying or beheading their enemies on the way there, the fighters of the Islamic State who've taken over the city have now banned cigarettes from the shops.

At least I think they have. Like much that we hear these days in Iraq, from the numbers of refugees still left on Mount Sinjar, to the fighting that's still very much going on, some things remain opaque, while others grow in the telling. But what's clear is that the Islamist militants' reputation for savagery is enough to drive hundreds of thousands of Iraqis and Syrians from their homes.

Mergeen tells me about Mosul – the city he grew up in – as we drive through the city of Dohuk. The Kurdish businessman and his family, like so many others, are looking after relatives who fled Mosul. But friends who remained say that ISIS is blowing up churches, and the mosques they don't approve of, all in the name of religion. 'We all lived together there, and nobody cared who was what,' he says sadly. That's no longer the case for much of Iraq today.

In the centre of Dohuk, with its cafés and busy shops, a humanitarian crisis has been quietly unfolding. It took a mountain filled with fleeing families to put it centre stage. But now the plight of the Yazidi refugees is all too visible. First, in their flight to the mountain, and now in the makeshift sanctuaries here, in derelict buildings, the men, women and children crammed together without water or food. Their exodus from the mountaintop and arrival on foot, the old and the young, in the heat and dust, mothers clutching their children, looked like an image from the Old Testament.

In Dohuk, though, they are far from the only ones fleeing the terror. This once-obscure religious minority has joined many other refugees, from Christians to Shabak, Kakai, Armenian and Turkmen minorities, as well as Shia Muslims. The Yazidi are just the newest of the one million displaced and scattered across northern Iraq by the fighting in Syria and Iraq. And it's left this province struggling, despite the shopping malls and the glossy new buildings that are the fruits of oil revenues.

The governor of the province, Farhad Atrushi, only took office a little while ago, but has the air of a man with many things on his mind. We meet at his hillside home on a balmy evening, overlooking the waters of the lake. He greets us in fluent English. We are not his only visitors; a German delegation has just left, then some Iraqi MPs arrive. All want to know how this humanitarian crisis can be resolved. And a crisis it most certainly is.

He knows they are working against the clock. It takes six weeks to build a refugee camp, he says, and many more are needed. But epidemics don't wait that long. And the money simply isn't there: Baghdad hasn't agreed a budget for this area since the spring. But thousands of children have arrived, many of them sick, almost all unvaccinated, from one of the country's poorest provinces.

I ask why the West should intervene with aid or arms, and his answer is persuasive. 'Because we face a calamity. We don't have long, and if we wait, more people will die. A third of Iraq is living in terror. We have to save the Yazidi – and protect what we have here: people who live together in peace, in the part of Iraq that's been a beacon of hope.'

And it's true. In places that aid agencies have not yet reached, local people have brought home-cooked food, fresh water, clean clothing and much more, to help families they don't even know – moved by a common humanity, and a remarkable tolerance of the dispossessed visited upon them.

To the west, and the south, the Islamist fighters are still advancing, trying to carve their way across Iraq, though Kurdish *peshmerga* fighters are now digging in and fighting back. ISIS has young men willing to die to impose their joyless vision. But all here insist the Kurds are also ready to defend the harvest they have sown over many difficult decades.

'If the West doesn't intervene, and quickly, the one bright spot in Iraq could be extinguished,' warns Farhad Atrushi. And, he says, if IS isn't stopped, a safe haven will grow from which to sow fresh terror. He has hope for the new Iraqi government: that even if it can't reconcile mistrustful ethnic and religious groups across the country, it might at least let them work together.

As we leave, a heavy crescent moon hangs over the governor's house. I can't quite tell if it's waxing or on the wane as it lights up the warm Iraqi night.

Orla Guerin – Back to Bombings, Birth of IS – 12/06/15

Trying to explain the speed and relentlessness of Islamic State's advance called for a longer view of how political and religiously inspired violence had become so prevalent. Many Iraqis were appalled, but not greatly surprised, by reports of their troops simply abandoning cities like Mosul after IS launched its big push. How could the Iraqi armed forces have crumpled so fast – and how far did the writ of central government in Baghdad really run? Orla Guerin had lived through previous jihadist offensives in Iraq, but was unsure how long the country could survive the sectarian tensions tearing at its cities and its people.

Two weeks ago, I was dropping off to sleep when history repeated itself with a sickening thud. It was a bomb, nearby, so near it shook the BBC bureau in Baghdad. Minutes later from the roof I could see dark smoke rising in front of an upmarket hotel – now called the Christal, but formerly the Sheraton. It was one of two hotels targeted that night by the Sunni extremists of Islamic State. Fifteen people were killed.

Standing on that rooftop, hearing sirens wail, brought back another assault – that one on Baghdad's Sheraton, in November 2003. Back then I was in the hotel, during a rocket attack. There was a direct hit on the sixteenth floor. It sent the lift crashing into the lobby but no lives were lost.

Baghdad looks a little better these days. There are fewer blast walls and checkpoints. When the searing heat of day fades, you can see families chatting on low walls along a section of the River Tigris. Those who can afford it sit on wooden benches eating grilled fish and puffing on water pipes. But any sense of normalcy here is easily shattered, like the countless shop fronts that have been smashed by bombs.

Iraq's newest enemy, Islamic State, is in residence in the city of Ramadi just seventy miles west. We came face to face with members of an IS cell in the capital, who are now in custody. They are accused of helping to plan bomb attacks that killed about fifty people late last year.

The cell leader, Haider Mansur, limped into view, in a yellow prison uniform. We were told he'd injured himself trying to evade arrest. The thirty-four-year-old was handcuffed and shackled. He said he was studying accountancy before Islamic State came calling; with his short hair and neatly-trimmed beard he almost looked the part. He calmly admitted helping to arrange eight bombings, though he said he avoided news reports about the attacks. 'I didn't want to know how many died,' he said, claiming he'd felt guilty.

One of his accomplices, Loauy Saleh, was a jobless father of four.

He was balding and softly spoken. He claimed he only joined IS because they bought him a car and promised to rent a house for his family. But he also said he was attracted by their promise of a region for marginalised Sunnis like him.

Islamic State feeds off the dissatisfaction of the Sunni minority. Mansur and his men are in some ways a symptom of a broader danger to Iraq: sectarian hatred. 'Once IS is defeated, our problems will increase,' said a seasoned Iraqi journalist. 'The sectarian issue is the biggest threat. The next war will be about that.'

It is already a familiar battle here. The country was riven by sectarian killings in 2006 and 2007. The situation is far more dangerous now, according to the journalist, himself a Sunni. 'I am fifty-two years old,' he said, in precise English, matching his neat attire. 'I have always lived in Baghdad. And this is the worst year of my life. I worry about Islamic State, but I also worry about being kidnapped by Shia militias. Every time I leave home I drive a different route.' Asked if he thought there was a future for his homeland, his response was swift. 'Not at all,' he said, his voice rich with sadness.

What will the coming years mean for Iraq? What kind of landscape will be carved out by Islamic State and sectarian power struggles?

I got one answer after coming within sight of IS territory in Northern Iraq. Kurdish fighters took us to the windswept front lines near the city of Mosul. We could see the plain of Nineveh down below us, a great expanse of fertile ground held by Islamic State. As we drove away our local guide spoke up. 'We need independent states for Kurds, Sunnis and Shias,' he said. 'Iraq is on life support. It's time to switch off the machine.'

Quentin Sommerville – The Cubs of the Caliphate – 13/02/16

Those lucky enough to survive or escape life under IS control had tales of horror to tell. Christians and Yazidis had often been forced to convert to Islam under threat of death; many women and girls had been held in sexual slavery, sometimes bought and sold many times to different

IS fighters; some captives had been brutalised and executed on camera. Foreign aid workers and journalists were among the victims, but many more were soldiers – Iraqi, Syrian and Jordanian – whose deaths had been carefully orchestrated to provoke maximum reaction from their home governments. But IS had also, as a matter of policy, traumatised its own members – and the children it exhorted them to raise. Quentin Sommerville heard chilling details of how the group had planned to indoctrinate its next generation.

Isn't it wrong for an adult to wish nightmares on children?

Mohammed is eight years old, has dark spiky hair and a gravelly, serious little voice. He sat in front of me and showed me, without thinking, the practised and efficient motions of putting on and detonating a suicide vest. 'You tie it tightly ... pull the straps here ... and the detonator is in this hand.' His small, fat fingers gripped tightly. 'Boom,' he said, as he pushed down the imaginary trigger. Beside him, his cousin Ahmed, aged ten, with big, sad blue eyes and neatly gelled blond curls, told me about the killings they'd witnessed.

Both these boys lived under the so-called Islamic State, were schooled in murders and executions, and trained for the day they would fight for the caliphate. It's the stuff of bad dreams, but they insisted that neither of them had nightmares. There they sat, wearing their new jeans and trainers – bought especially by their uncle for our TV interview – and I thought: what damage has been done here?

They'd be called on loudspeaker to watch beheadings. Saw two of their neighbours killed. A close relative was maimed. 'Once a fat man came and told the prisoners they'd be shot, not beheaded,' Mohammed recalled. Two men were killed in front of them, in the middle of the road. Afterwards they were beheaded. 'They put their heads on spikes at a roundabout,' the eight year old told me.

The boys were schooled in sharia law ... and how to load a Kalashnikov. 'We were allowed to fire one bullet a day,' Ahmed said. If anyone got scared, they were given sweets. For treats, Islamic State fighters would take the class swimming.

IS is attempting to radicalise a generation. The group is planning for its fantasy future. The boys were told they were cubs of the caliphate; they were promised they would become fighters, martyrs. And that they would kill *kuffars*, or infidels.

I spoke remotely to a schoolteacher still inside IS-held Syria. Lessons and books are being rewritten, he explained, history and geography altered in an attempt to create an IS version of the future. The Ottoman Empire had been a good thing, IS books insist. The caliphate, instead of being a shrinking, broken territory that straddles parts of Iraq and Syria, stretches from the Atlantic to China in their lessons. But the teacher explained that only a fraction of schools are operating, and staff haven't been paid in months. If this is a state, it appears to be one that barely functions.

Ahmed and Mohammed were taught to hate Shia Muslims especially. Mohammed explained: 'They said Shias don't pray and that they kill those Muslims who pray in our mosques. They said Shias don't believe in God – that they believe in a person and not God. They believe in Bashar Assad.'

Children also serve another purpose for IS. The cubs of the caliphate appear in propaganda videos, carrying out unspeakable acts. A British child, named as Isa Dare, has appeared in them twice. In another video released this week, a British child detonates a car bomb, a vehicle with prisoners of IS trapped inside it. The boy's grandfather says he's a pawn and being used as a human shield.

When Ahmed and Mohammed were finished with their interview, they joined their younger cousin and fought and squabbled over a mobile phone game. Suddenly they were children again. Physically they're safe now, and living in Turkey with their uncle, but the damage has been done.

Later, I was in a refugee camp on the Turkish border where some Syrian kids were running around playing with toy guns. One boy showed me the model AK-47 his brother had made him. It was well put together, with a carved ammo clip, a screw for a trigger and a glue nozzle for the gun sight. He and his pals were playing war games. Someone got hit on the head . . . and went home crying to his mum. It was just a game; the weapons weren't real.

As they laughed and fought, I noticed one of the kids looked just like little serious Mohammed. I hoped that when they all went home and their parents tucked them into bed at night, there would be no nightmares for them.

Yolande Knell – IS Trials 02/09/17

By 2017 US-backed Iraqi forces and the Kurdish militia were regaining key territory from IS, retaking Mosul in July and the militants' strong-hold of Tal Afar, near the Syrian border in September. But efforts to seek justice for the many victims of IS were seen as less of a priority than eliminating the group. In the badly damaged, mainly Christian town of Qaraqosh, twenty miles south-east of Mosul, suspected IS fighters and collaborators were put on trial in two dedicated courts. Yolande Knell sat in on the hearings.

The young man wears a shabby, brown prisoner's outfit and his legs are shaking nervously as he stands before the three judges robed in black. But after sipping from a glass of water, he confirms his name: Abdullah Hussein. He's accused of fighting for the so-called Islamic State.

'The decision of the court has been taken according to articles 2 and 3 of the 2005 Counterterrorism Law,' one judge states formally. 'Death by hanging . . .' And then Hussein, who like many suspects here was picked up on the Mosul front line, breaks down crying.

He's standing just an arm's length from me in a tiny room in Qaraqosh, where the provincial courthouse now hears criminal cases for the whole Nineveh region around Mosul. For several months, it's been trying IS-related cases. I'm told there are some 3,000 of them to get through.

The next defendant, Khalil Hamada, is twenty-one and more forthcoming. His town was held by IS for two years and he recalls seeking out its local recruiter.

'I went by myself, nobody forced me. A lot of us joined,' he says.

'How did you join? What oath did you take?' the judge asks.

'I can't remember exactly,' Hamada replies. 'But I swore loyalty to Abu Bakr al-Baghdadi and the caliphate.'

He goes on to recount how he underwent training with IS in sharia law, bodybuilding and using weapons. But he tells the court he became 'just a cook' – before admitting that he was also a guard, at an IS base. He was paid £120 a month. When the judge summarises his story, Hamada nods: 'Yes, it's true.'

A woman prosecutor speaks and then – albeit briefly – a state-appointed defence lawyer. During the judges' brief deliberations, I stand in the corridor by a line of handcuffed prisoners squatting on the floor.

Khalil Hamada is also sentenced to death. Like all those convicted, he's told he can appeal and that a higher court in Baghdad makes final rulings. But his look of resignation suggests he believes this is a formality. International human rights groups have raised concerns about the ill treatment of IS suspects and the fairness of these trials; there are as many as thirty of them held each day. But the judges here insist they act professionally and are sending a strong message that the Iraqi government is restoring law and order.

'The judge remains neutral,' says Justice Younis Jameeli, reminding me that IS fighters singled out the Iraqi judiciary in Mosul, killing fifteen of his colleagues. 'Each of us lost family members and had homes destroyed but when a suspect appears before us, we treat him according to the law,' he goes on. 'That's the judicial process.'

This judge – a lean man with a trimmed moustache – heads the Investigations Court, which is set up for now in a large family home. His office is a bedroom, the dressing table replaced by a desk piled high with documents. When I ask about evidence, there's a glint in the judge's eye. 'You know IS themselves are helping us convict them,' he declares, reaching for a file. 'Look here ...'

Inside there's further evidence that IS aren't a disorderly militia but meant to function as a state. It's a spreadsheet from a computer recovered by Iraqi intelligence. Each of the 196 rows neatly identifies an IS member – his full name and address, his job and a photo.

With real fears that jihadists will try to blend back into the

Iraqi population, prosecuting them is meant to try and stop IS re-emerging as an insurgent group. And also to prevent reprisals being taken *outside* the legal system. Near the court, I meet Muwafaq, who's come from Mosul to make an enquiry. He tells me his neighbour, who joined IS, burnt down his home. 'I hope he gets to court before I see him!' he says menacingly.

But others at the hearings allege their loved ones were wrongly arrested. One woman claims her husband has mental health problems. One father insists his son was 'a regular guy selling vegetables from a cart', not part of IS. Talking to them, it's clear that judging exactly who was a collaborator is a tricky business. It's hard to tell whether some locals did what they had to, just to survive, or whether they really bought into the extremist ideology.

Around us, the largely deserted streets of Qaraqosh and its desecrated churches attest to IS's methods. Three years ago, thousands of people fled this mostly Christian town as the jihadists advanced. So far, very few have felt it's safe to return.

As court proceedings close, armed guards march a column of prisoners past us out of the gate. They have their heads down, looking at the dust.

Part IV:
SOUTH ASIA

Afghanistan

Hugh Sykes – Cinema Park, Kabul – 24/07/10

The American intervention in Afghanistan, aka 'Operation Enduring Freedom', certainly did endure – officially, from 2001 to 2014. There were many warnings that the country was a morass: no foreign force had ever been able to take it, let alone hold it. Genuine progress was made: there were free, multi-party elections in 2004, when Hamid Karzai became president; many women returned to work and educa-tion; a generation of Afghan exiles began to wonder whether it might be possible to return. But the Taliban never went away, and there were deep concerns about the legitimacy of whoever might replace them.

Under Taliban rule, notoriously, many places where Afghans had once gone to enjoy themselves were either closed down or turned into gruesome showcases for the theatre of punishment. Football stadiums became venues for executions, amputations and stonings. Images of living things were destroyed or defaced, while moving images were particularly abominated. Hugh Sykes stumbled upon a corner of Kabul saturated with memories, and heard of plot twists yet to come in Af-ghanistan's endless national epic.

I've just been to a place that reminded me of that lovely Italian film *Cinema Paradiso* . . . the clatter of the projector, the faint image of the big screen seen through the little window of the projection booth, whistles and catcalls when the sound fails.

But this is Cinema Park, Kabul. Five hundred and forty-four seats. Indian films mostly – and James Bond, and Rambo. Only the circle is open, with a low concrete wall across the front instead of a guard rail. And plush upholstered flip-up seats . . . very dusty seats. As the beam from the projector cuts through the dark, it catches strands of cigarette smoke curling up from the dozen or so young men who've come to see an Indian action film. The sound is distorted, and the 'print' so scratched that often images are almost obliterated. The stalls are locked and unused. But all the old seats are still there – even more dusty, hard, flip-up wooden seats, with numbers stencilled on the back.

Halilullah is the manager. He started working at Cinema Park thirty-two years ago. He and most of the staff fled when the Taliban came in 1996. They saw them coming, literally – nine turbaned, bearded Talibs with guns over their shoulders walking through the pine trees in the park next to the cinema. Not all the staff got away in time. The Talibs beat them and said they were pimps for Western decadence, but they let them go. They also smashed the old projectors, and camped in one of the offices.

When Kabul was liberated in 2001, Halilullah came back. He doesn't intend to leave again. And he doesn't think the Taliban will come back. But many people here are afraid they might. The pessimists see a menacing future, in which Western countries succumb to sentiment (pressure from soldiers' families) and to misguided self-interest (fear of losing mid-term elections) and bring their troops home before the mission is complete. Afghans are also deeply alarmed that President Karzai recently went on television and said '*Taliban-jan*' – which means 'dear Taliban'.

Deadly enemies of the Taliban, such as the Northern Alliance, are profoundly unsettled by reports that Hamid Karzai has already held talks with the leader of the Haqqani network, the best-funded and most ruthless Taliban group based in the tribal areas of Pakistan.

And then there's corruption. There were pledges at the Kabul conference that corruption would be tackled. But it is endemic here at every level. A survey by the United Nations Office on Drugs and Crime found that half the people questioned admitted to having paid a bribe for a public service: 25 per cent had bribed a policeman, 18 per cent a judge and 13 per cent a lawyer.

But there is also foreign corruption.

In a gaudy new Kabul shopping mall, I was having a very good espresso coffee when the man at the next table invited me to join him. Abdullah is a hard-smoking Afghan reporter who's spent the past eighteen years in Moscow. This is his first time back here. And when I ask him an open question about his impressions after eighteen years, the first thing he says is, 'I've heard the promises of billions of dollars for Afghanistan but when I look around this city I don't see much evidence of it. A few new buildings, a few new roads since I was last here, but not much else.' And, he goes on, 'I think I know where the money has gone: into the British and American mafia – the mafia of foreign contractors and all the foreigners here on inflated salaries. By the time the money reaches Afghans, there's hardly any left.'

Abdullah has a point. Foreign civilian advisers can earn $500 or more a day here ($500 is about three times the average *monthly* Afghan wage) and then there's the cost of their expensive secure accommodation and fuel-hungry four-by-four vehicles.

The people in the streets around Cinema Park – near those pine trees where the Taliban walked before they smashed the projectors – could do with some of this foreign cash. To fix the roadside gulleys full of foul, stagnant mosquito-infested black water, or to help buy clothes for the boys in rags and no shoes who beg on the cinema steps.

Andrew North – Waiting for the Boulder to Fall – 27/10/12

While there was hope of avoiding a return to all-out civil war, the people of Afghanistan could never relax. Many former warlords were still at large, commanding huge private armies and untold profits from corruption and narcotics. (Even the brutal prohibitions of the Taliban hadn't rooted out opium-poppy farming, and national heroin production boomed during the decade after they were removed from power.) The country was still full of well-armed men with grudges, and while the Taliban insurgency surged and ebbed, it never seemed to end; bombings and assassinations came all too regularly. Andrew North found many of his friends holding their breath to see what might come next.

It wasn't long before 2014 came up. I was having a meal with a group of Afghan friends, sitting cross-legged on floor-cushions and tearing into a plate of spiced lamb kebabs. We had heaps of flatbread, onion yogurt and giant Afghan radishes – each the size of a doorknob, the best you'll taste anywhere.

'Do you think you could get me a visa?' said one friend I'll call Saleem, as he paused to sip a glass of green tea.

In different forms you keep hearing the same conversation as the twin deadlines of NATO's pull-out and Afghanistan's next elections draw nearer. The approach of 2014 looms in everyone's mind like a boulder balanced precariously on a mountainside, threatening to come crashing down. Some are already moving before it can roll, fearing a new civil war, quietly getting their families abroad. As I talked with another friend about someone we both know returning to Afghanistan after studying in Europe, he laughed in surprise: 'Why did he come back?'

The government of President Hamid Karzai angrily attacks such doom-laden talk. 'Nonsense' and 'garbage', officials said, about a recent international think-tank report predicting a post-2014 meltdown. But as we passed round the last of the kebabs, soaking up the juice with hunks of bread, the blame for the pessimistic mood was being shared out equally between Mr Karzai and the West.

A friend who works with foreign contractors was telling tales of their profligacy, spending huge sums on renting accommodation that was never used – all paid for by Western taxpayers. But he also pointed the finger at the Afghan leader, for his actions, or lack of them.

So far he's done nothing to allay fears of a repeat of the systematic fraud in the last election. There are concerns Mr Karzai will try to fix the result in 2014; not for himself, as the constitution prevents him standing again, but for a proxy who he can control. There is speculation that proxy could even be his brother.

Just smears, Mr Karzai's backers say. Afghanistan is looking forward to an election free from foreign interference, was how one of his advisers put it. 'Of course they are,' said a prominent Afghan politician I was talking to, who used to support Mr Karzai, 'so this time they can do whatever they like.'

His sourness is a flavour of the tension that some fear will spill over even before the election – toppling that boulder and sparking a renewed battle for Kabul. At least 50,000 people died in the city during the 1992–96 civil war – far more than during the past decade – and the faction commanders responsible for that carnage have used the last ten years to regain power, untouched by any process of justice. Backed by past American ambassadors, President Karzai decided that letting them be was the safer course. Yet to many Afghans, it looks like the government has gone further and is trying to rehabilitate them.

On my way to see a doctor friend the other day, I passed one of Kabul's universities. Seeing the big freshly painted sign above the entrance, my taxi driver shook his head in disgust. The sign read 'Burhanuddin Rabbani' – preceded by the word for 'martyr'.

A small tribute, you might say, to the man who headed a peace council charged with trying to reach out to the Taliban – until he was assassinated, probably by the Taliban. But Kabulis remember him better as the president during the civil war, whose own faction helped turn the city to rubble. The decision to paint his name provoked days of clashes between university students and police.

As America's longest war enters its twelfth year, there's cynicism and suspicion all round. While NATO soldiers now fear being shot in the back when they venture out with Afghan troops, it was telling that when the British Ministry of Defence recently announced the arrest of several Royal Marines for allegedly murdering an insurgent, the Afghan media simply assumed that they'd killed unarmed civilians.

This distrust is made worse by the fact that most foreigners here are more separated from Afghans than ever. Not just the military, but diplomats, aid workers and even journalists too – living behind higher blast walls, never walking, only venturing out in four-wheel drives. 'Somehow they still manage to write books about Afghanistan, though!' Saleem laughed sardonically.

Yet for all this gloom, the worst predictions may still turn out to be wrong – for the simple reason that many more people have a lot more to lose now. When the civil war began in 1992, the commanders were in effect fighting for the spoils after the Soviet withdrawal. Now, after ten years of NATO intervention, they are wealthy and many hold positions in government.

While there are fears the Taliban will regain control in their heartlands in the south, it'll be much harder to repeat their takeover of the whole country, which they achieved partly by buying people off. The cost of that will now be too high. After all the blood and treasure that's been spent here, this is hardly a verdict of success – but it may be enough to stop that boulder falling.

Kate Clark – Twenty Years On – 13/10/16

Outside analysts and reporters – let alone would-be interventionists – were often exhorted to take the long view in Afghanistan: to learn the proper lessons from history, or be doomed to repeat them. Kate Clark was the BBC Kabul correspondent and the only Western reporter in Afghanistan in the last years of the Taliban; she now writes for a research organisation in Kabul, the Afghanistan Analysts Network. She reflected on what had really changed after nearly two decades of war.

When I first came to Kabul in 1999, a third of the city lay in rubble. Buildings left standing were pockmarked by shrapnel and bullets. I didn't meet anyone who hadn't lost a brother, a mother, a son in the brutal civil war of the mid 1990s. The Taliban had put a stop to that fighting when they'd captured Kabul three years earlier. They brought order and security to the city, but also imposed what to Kabulis was a strange vision of Islam. Behaviour that was normal where the Taliban came from – the rural Pashtun south – was now being imposed in the capital by the religious police, with whips and threats.

Speaking out publicly against the new regime was very dangerous, but in private there were tales to tell. One woman, a former teacher, said it was 'like living in prison'. Her husband spoke of his despair at seeing his daughters growing up uneducated. Widows without male support grew hungry and desperate.

There was also resistance. Kabuli women were forced to wear the burka, but walked with their heads held defiantly high. Girls were educated secretly at home – or not so secretly. I visited one fully functioning school with blackboards and classrooms. The Taliban usually turned a blind eye to the clandestine education of younger girls, although teachers always ran the risk of punishment.

Kabul bustled even then, but with pedestrians and bicycles. Rush hour was a massing of tens of thousands of bikes. You'd also see male cyclists transporting women balanced side-saddle, clutching the cloth of their burqas out of the wheels.

Poverty was rife – but then no one was particularly rich. The Taliban, for all their faults, were not particularly interested in making money. With few cars and little electricity, Kabul had clean air then – you could see the stars and the jagged mountains that circle and intersect the city.

Then one night in November 2001, after weeks of US bombing, the Taliban fled and a new era began. Mammon suddenly seemed to be in charge. There were huge, gushing flows – then floods – of foreign money, both aid and military spending. Kabul grew. Near where I live, the chief of police bulldozed people's mud-built homes in 2003, clearing land for the wealthy and well-connected to put up huge, gaudily painted mansions. Many call them 'poppy palaces' because of the narco-dollars assumed to have paid for them.

As incomes rose, the city became clogged with traffic. A brown pall of pollution now hangs over Kabul. Some Afghans barely survive, while others have made millions, even billions, out of the US-led intervention – getting eye-watering contracts, or by siphoning off state revenues, grabbing land or living off the booming opium trade. There are a few true entrepreneurs, but not many.

The population of this city has quadrupled, quintupled, maybe more – no one is quite sure how many people now live here – as refugees returned, and people fled rural poverty and a new phase of Afghanistan's long and bitter war. This time, it's largely being fought out in the countryside. But it doesn't necessarily stay there.

Recently, Kunduz in the north, and Helmand in the south – where British troops were posted – have seen Taliban moving in from the districts and entering the provincial capitals, Kunduz city and Lashkar Gah. The reasons behind the government's inability to protect its people are pretty much the same in both places: infighting and state corruption. Last year it was found that two thirds of the Afghan soldiers being deployed to Kunduz were 'ghosts' – mere names on paper, whose salaries were being pocketed by their superiors. Not long ago, the Afghan Army general in charge of the province was accused of selling rations to his own troops.

You couldn't call Kabul exactly safe either. But here it's the occasional bomb or suicide attack, rather than rocket fire or sustained fighting, that disturbs the peace – and life carries on almost as normal. Afghan girls continue to go to school; whenever I see them spilling out onto the streets after class, chatting excitedly as they walk, it makes me happy. Women can work now, although patriarchal expectations still stunt the lives of many. The struggle continues against early marriage and violent husbands, and to be allowed to work or get an education. But at least it's now negotiated within the family or community. Afghan women are still riding in the back, though now they're in cars, not on bikes. In 2016, only a handful of them are in the driving seat.

Auliya Atrafi – The Pragmatic Taliban – 08/06/17

Perhaps the most bitter pill for British observers to swallow was watching the Taliban take back towns and villages in Helmand Province, which Western forces had fought for at great human and military cost. The place names which had once featured regularly on British news – Sangin, Kajaki, Lashkar Gah – were back in the bulletins, but now with the Taliban retaking control. They'd seemed to triumph through sheer persistence, but had their tactics changed? As an Afghan, a Muslim and a son of the province, Auliya Atrafi was better placed than many journalists to meet the fighters who'd recaptured the territory, and gauge what they really had planned for its people. He also uncovered a surprising 'dual strategy' of simultaneously attacking the central government in Kabul and pursuing limited cooperation with it.

The Sangin evening seemed a particularly pleasant one . . . or maybe it was simply relief I was feeling.

There were twenty or so of them, sitting cross-legged inside a huge mud compound. Under the light of the full moon, their black turbans cast deep shadows over their sunburned features. I took a tentative sip of the green tea they had given me.

These were the Taliban's Special Forces: the Red Unit. They sat quietly as they listened to their commander Mullah Taqqi telling war stories. They gently rocked their M4 machine guns on their laps, as if sending babies to sleep. The M4s – with their night vision scopes – were treated with the same care as newborns. After all, these weapons were the secret of the Taliban's success, the reason they had captured nearly 85 per cent of Helmand Province.

But these fighters' military victories had presented their leaders with an unexpected challenge. The people they now ruled had lived with central government services for more than a decade. Schools, hospitals, development – all these had become expected here. How could an organisation dedicated to taking territory adapt into one that could run it? It was this I had come to see.

Our first encounter with Taliban governance came the next morning. The old Sangin bazaar had been flattened in the battle for the city. We were walking through its makeshift replacement, a sea of tarpaulin and boxes. Two men were arguing by a food stall over a box of apparently expired biscuits. The one who appeared to be the shopkeeper was called Haji Saifullah; he was agitated. 'I can't read!' he shouted. 'How was I supposed to know it was out of date?' He fidgeted with his turban, pushing it to one side nervously.

The other man turned out to be Sangin's Taliban mayor, Noor Mohammad. He had already imprisoned Saifullah for three days; now he was sending him to the district court to be fined.

Experiences like that of the illiterate shopkeeper have made rural Afghans realise that education will not turn you into an infidel, as their forefathers feared. Rather, they are starting to see it as essential.

Next on the mayor's list was inspecting petrol jugs to see if they'd been altered to pour less than the promised gallon. Then on to holding medical exams for people who claimed to be doctors and pharmacists, but whom he suspected were lying. This was health and safety, Taliban style.

If Taliban concern over food standards had been surprising, what I discovered in Musa Qala took my astonishment to another level. Despite being in the Taliban capital, the school and hospital were still being funded by the government in Kabul. The Afghan government was paying the salaries and providing the equipment; the Taliban were administering the services. The syllabus was set by central government, but was taught in Taliban territory. The Taliban were trying to present themselves as a legitimate government, but Kabul could maintain influence in an area, even though it had no physical presence.

There were other surprises. The internet is banned in Musa Qala for security and religious reasons ... but we were told there are dozens of Wi-Fi hotspots connected to the outside world. A few dedicated fans of Turkish and Indian soap operas still manage to watch them on televisions hooked to small satellite dishes. 'Aren't you scared the Taliban will find out?' I asked one mischievous-looking teenager. 'They know about our TV and the Wi-Fi,' he tells me. 'It would be naive to assume they didn't. But I think they are just watching and waiting, to see what happens.'

That thought took me back to my night sitting drinking tea with Mullah Taqqi. Despite his fearsome reputation, he had a thought-ful, almost gentle, air. We discussed politics, government, love, life and death. It struck me that the Taliban here were experimenting, tentatively, with the ideas of modernity. They realised they could not fight the modern world forever, and so had opted to join it – but on their own terms.

I took a long sip from my cooling tea. It made me think: perhaps this was the lasting impact of the British and American intervention in my country, as well as the central government. Yes, the foreign forces were gone, but they had left something unexpected behind: a feeling in the people that their rulers are not simply there to make rules but to provide services as well. The outsiders have changed ordinary people's expectations and, in doing so, they have challenged at least some elements of the Taliban to do the same.

Pakistan

Mishal Husain – Abbottabad – 07/05/11

'I can report to the American people and to the world, the United States has conducted an operation that killed Osama bin Laden, the leader of al-Qaeda, and a terrorist who's responsible for the murder of thousands of innocent men, women and children.' On 2 May 2011, Barack Obama finally announced the end of a mission which obsessed America and the world for over a decade. But the death of Osama bin Laden provoked many questions. How could he have hidden in Pakistan – supposedly a country allied to the US – for so long? In a press conference, the head of Pakistan's foreign ministry dismissed as 'absolutely wrong' any specu-lation that bin Laden must have been shielded, in his hideaway in the town of Abbottabad, by sections of the Pakistani security establishment. Mishal Husain went there to hear what bin Laden's unwitting neigh-bours thought.

My first sight of the house that had been Osama bin Laden's final home came after we drove down a narrow dirt road, winding through a neighbourhood of newly built and relatively affluent homes. As the road cleared into an open ploughed field, with just a few houses dotted around, there was no mistaking that we had reached our destination. The satellite vans were parked back to back, and crowds had gathered around the perimeter walls of the compound and in front of the imposing green gates.

I stood in front of the building and tried to imagine if there was anything that would have made it stand out to me, if I had passed by *before* bin Laden made it so notorious. The walls were slightly higher than those of the neighbours, and there were a few lines of nasty-looking barbed wire at the top. The grounds were also more substantial than the houses nearby – clearly this establishment belonged to someone affluent enough to dedicate a large proportion of their land to a garden.

Among the crowds gathered outside was a group of young women who said they came past here every day on their evening walk. What did they think of the news that it was home to America's most wanted man? 'I don't believe it,' said one. 'Osama bin Laden was just too important a man to be living here in Abbottabad.' Didn't they think there was something odd about this fortress-like home, I wondered? 'No,' said another. 'With all the bomb blasts in this country, plenty of people live behind walls like this.'

On the night of the American raid, they told me, the sound of the helicopters was heard for miles around. Many assumed it was a night-time Pakistan army exercise, but then the firing began. Some told us they had seen burning debris on the ground, but by the time the area around the compound was opened to us, it had been well cleared. There was no sign of life inside – but a heavy police presence at the gates.

As we headed back to our hotel, I passed the main entrance of a very different Abbottabad establishment – one whose name I have known all my life. Burn Hall was my father's school in the early 1950s, when he came here to be educated by Christian missionaries. It's still there today, although it is now run by the army and educates girls too.

Stories about Burn Hall and the charismatic, if slightly forbidding, Catholic Fathers who taught there were a feature of my own childhood, especially on our annual fishing expeditions to the north of Pakistan, when we would invariably pass through Abbottabad. This was where we stocked up on essentials, packing a few more supplies before we began the ascent into the mountains and then into the valley that was our regular summer haunt.

Abottabad always struck me as a very un-Pakistani sounding name, until I realised it was actually named for a certain James Abbott, a British officer who became a near-legend in this region in the 1840s and 50s, and was knighted for his efforts to bring it firmly under the control of the Raj.

Curiously, although many pre-independence, British-sounding place names in Pakistan were changed over the years, this reminder of the colonial past remained. I'm sure few of Abbottabad's current residents know much about General Abbott though, and it's now Osama bin Laden who has put their town on the map.

Close links to the Pakistan army are very evident here, from the roadside tank that alerts you to the nearby military academy, to the monument that proudly recreates and pays homage to one of the country's nuclear missiles. In the shadow of this giant weapon-shaped sculpture, I encountered more disbelief about recent events. 'The Americans must prove to us that bin Laden was here!' said one man, proprietor of a small grocery store that charmingly called itself a 'tuck shop'. His friend agreed: 'If they can really come here, catch him and take him away, all in the space of half an hour, surely they can show us the body?'

But those doubts don't mean that people here have any great sympathy for Osama bin Laden. Almost a decade after the 9/11 attacks, many Pakistanis feel that they have paid a high price for their country's alliance with the United States – both in the tense security situation and in the freefall of their economy.

Today, I cannot pack my own three children into the back of a car and drive into the mountains of northern Pakistan as my parents did so freely with me.

But I'm looking forward to the day when I *can*, once again, make that journey. Though when we pass through Abottabad, I'll have to tell the children that it now has a place in history as well as one in the story of our family.

Shaimaa Khalil – Small Coffins Are the Heaviest – 20/12/14

The death of Osama bin Laden certainly didn't mean an end to jihadist violence in either Afghanistan or Pakistan. Although the Tehrik-I-Taliban of Pakistan share Pashtun heritage and a hard-line version of Sunni Islam with their counterparts in the Afghan Taliban, and both movements consider themselves national liberators, their tactics and loyalties differ. As security analysts and agencies tried to unravel the intricate relationships between the two groups – and what remained of the al-Qaeda network – civilians kept dying in attacks on both sides of the border. As the Pakistani military (with support from US drones) picked off more and more of the TTP's leadership, the backlash was merciless. In 2014, Shaimaa Khalil saw the aftermath of its most lethal attack to date, in which 141 people were killed at the Army Public School in Peshawar. Nine of the dead were staff members; the rest were children aged between eight and eighteen.

You didn't need to see the pools of blood in every corner of the school auditorium to know that a massacre had taken place there. A thick smell hit you as you went in; the eerie silence of the shocked adults, looking at the remnants of the children's belongings strewn around, itself told a devastating story. Almost every row of seats in the hall held similar pictures of horror: the blood-splattered textbooks, the school shoes left behind in desperate attempts to escape, scattered bits of paper on which you could see the children's handwriting . . .

This is where Taliban gunmen stormed in and shot over a hundred pupils at close range. The students were in here learning about first aid when the attack happened. One corner of the room bears scorch marks: that's where the militants set fire to one of the teachers.

When I arrived at the Lady Reading Hospital, where the surviving victims were taken, I could see a group of parents and family members gathering in front of a wall, pushing to get to the front. A list of names had been stuck up there – the dead and injured. As the day went on the list grew longer and the crowd around it bigger – all of them desperate to find out what had happened to their loved ones.

In the hospital's intensive care unit, I found thirteen-year-old Saeed lying on a bed, his family next to him. He was still too shocked to talk about what had happened. His mother Naheda is a teacher at the school. She told me Saeed had hidden under a chair and called her as the attack began. 'I heard everything,' she told me. 'My son was on the phone. He said, "I'm shot, please come and get me." And I didn't know how to reach him. As a mother, I can't describe to you what I felt,' she told me through her tears, adding: '*Alhamdulillah*, thank God, he's now here with me.'

But many other parents must now live with the fact their loved ones are not coming back. Like the Awan family, whose fourteen-year-old son Abdullah was shot in the face. At the family courtyard the female relatives gathered around the boy's corpse, which was lying in an open coffin. His slim body was covered in flowers; I could clearly see the bullet marks on his face. His mother sat on the floor weeping, wiping his forehead, holding his hand, still calling out his name. This was a family gripped by grief – wailing relatives, small children looking on in disbelief.

At the local graveyard the men recited verses from the Quran and prayed for Abdullah before laying him to rest. 'He was a top student,' his uncle told me. 'He loved birds, he had a couple of them at home and one of them laid eggs. He was counting the days for them to hatch but now he's gone.'

After the funeral I talked to the boy's father about his loss. 'He's with Allah,' the man said. 'And I have to be patient. I don't feel anything now. All these people are around me, they are supporting me. But,' he added, 'it will hit me tonight. When I'm in bed all alone.'

In the days after the attack the crowds gathering at the school gate grew larger and larger. The gates began to resemble a shrine, bedecked with roses, marigolds, candles and placards. There were messages of grief and solidarity. 'The smallest coffins are the heaviest' one sign said. 'My boy was my dream and my dream was shot dead' said another.

A group of teenagers from the army public school gathered there. They came dressed in their uniforms. Aakif Azeem, one of the senior students, still had some spots of blood on his jacket. He had survived the attack but told me how he'd seen many classmates die in front of him. 'I've been going from one funeral to another,' he said. 'I haven't slept for two days.'

Could he face going back to school after what had happened, I wondered? 'That's why I came here today dressed in my uniform,' he told me. 'It's a message to those who carried out the attack. You can take my friends. You can take my teachers. But I am still here. In my uniform. And I *will* go back to school.'

Shahzeb Jillani – Gwadar, China's Showcase – 22/06/17

Strategic analysis of Pakistan's future often centres on its historic rivalry with India – as both nations have nuclear weapons and tensions run so high over Kashmir, religious minorities and border security. Security specialists consider regional jihadi groups an even more immediate threat. But had the experts lost sight of an even more long-term trend? Just as in the rest of the world, Chinese influence is growing in Pakistan. Since mid 2005 there's been much talk of the Chinese foreign policy to set up a 'string of pearls' (that is, a network of Chinese shipping routes, ports and facilities) in key sites across the Indian Ocean. Shahzeb Jillani wondered what benefits such a grand scheme might bring to Pakistan.

The view of the Arabian Sea from Gwadar's only five-star hotel, located on the Koh-e-Batil Hill, is stunning. The building is shaped like a ship, facing the fast-expanding city beneath and the orange-coloured cranes of the Gwadar Port.

A few years ago, this hotel had to close because there were hardly any visitors. But nowadays there's a buzz in the air. The announcement of the China Pakistan Economic Corridor two years ago revived hopes for Gwadar, and with it the hotel's fortunes. Bookings shot up as government and business leaders started arriving to discuss the future of this once-forgotten fishing village. A sense of euphoria developed as Chinese businessmen and Pakistani property dealers signed multimillion-dollar deals.

From the luxurious, high-security surroundings of this hotel, the promise of Gwadar as China's new gateway to global trade appears very real. But take a drive down the hill to the city centre and it becomes clear why many locals remain sceptical.

'They are building new roads,' Mr Saleem, a fish trader along the scenic West Bay tells me. 'They say this place will become like Dubai or Europe – but what's in it for us? We are living the way we always have: without drinking water, electricity or fuel.'

The old town, with its narrow dirt streets, overflowing sewerage and heaps of rubbish, looks like any other neglected town in Balochistan. A dysfunctional municipal system and lack of basic social services make life miserable for many here. Clearly, for most of the 250,000 people of Gwadar, the fruits of progress have yet to arrive. So far, they appear to be silent spectators, sidelined by the changes taking place around them.

What they have seen over the last few years is the increased militarisation of their city. Pakistani security forces have acquired a firm grip of the shoreline. Navy and coastguard boats now strictly regulate the movement of local fishermen.

Abbas Waara is one of Gwadar's oldest boatmakers. He was a boy when he started helping out his family. 'This is all we have ever known,' he tells me. 'But now it's all going to change.' He's not against development, he says. 'I want my grandchildren to have the opportunities I never had – to get an education, to have a future.' But Mr Waara also believes that the local community doesn't figure much in the government's grand plan. 'I suspect one of these days we will be asked to move out to make room for new roads and construction,' he says with a sigh. Would there be relocation and rehabilitation for them? Mr Waara says no one knows yet.

The uncertainty is breeding political anxiety. Ethnic Baloch nationalists have long been resentful of the Pakistani army's control of the province. Now they fear that an influx of outsiders is turning them into a minority in their own land. Federal authorities in Pakistan are generally dismissive of such criticism. They accuse the nay-sayers of playing into the hands of India. And it's true that the Indian government worries about China's expanding influence in Pakistan.

Undeterred by criticism, China is pressing ahead. In Gwadar, its main priority is to revive the deep-sea port. So far, $160 million has been spent on it. An additional $500 million is being invested in a

Free Trade Zone. About 300 Chinese staff are already living and working there. But the port doesn't yet see much action – only a ship or two each month. Officials hope that by next year, that will go up to about three ships a week.

For Pakistan's civil military leaders, Chinese investment is a 'game-changer' for the country and the region. But for many people here, there's really one thing they need: access to clean drinking water. If the planned expansion does happen, Gwadar would need at least three times as much water as it currently draws from a nearby reservoir.

Over the years, the Pakistani government has spent billions trying to find a solution to the town's chronic water crisis. But those projects have suffered from mismanagement and corruption; the funds ran dry. To many, that says a lot about the government's capacity to realise its plans for Pakistan's future 'smart port city' and build a new Gwadar.

Secunder Kermani – *The Haircut that Crosses Continents – 27/05/17*

From the political loyalties encoded in Turkish moustaches to the nu-ances of a fringe peeking out of a 'proper' veil in Iran, hair can reveal an astonishing amount about a country. But few correspondents could have teased out quite as much insight into Pakistan (or the UK) as Secunder Kermani did from a haircut. It turned out that his own fa-voured style embodied a complex story of emigration, admiration and imitation between the two nations.

As I waited outside the blast-proof walls of my hotel in Islamabad, a short, moustached security guard wandered over.

'Are you from the UK?' he asked in Urdu. I told him I was. My family is originally from Pakistan, but I was born in London.

'So you're from Mirpur, are you?' he asked. I smiled: the city of Mirpur is known as 'little England' here, so many of its inhabitants left for Britain, only to return to build big houses with the money they've saved, or to marry off their children. No, my family aren't from there, I told him.

'Oh,' he said, looking rather disappointed whilst pointing at my head. 'It was your hairstyle – all the Mirpuris have the same one.'

To explain: I have what's called a skin fade, with the hair at the sides of my head shaved to varying degrees. Down by my ear, there's barely any hair, then it gradually gets longer as you go upwards. At the top of my head it's a normal, longer length. It's a very common cut in parts of Britain, particularly amongst British Asians. In Pakistan, though, most men have much simpler hairstyles.

I wasn't sure if the security guard liked my hair, but at least, I thought, I now knew where I could find a good barber. My first haircut in Pakistan would be in Mirpur, I resolved.

I was rather disappointed by Mirpur at first: it looked like most other Pakistani cities I had been to. But as we drove through the streets, I realised it was true: pretty much every other guy there under the age of forty seemed to be sporting a skin fade. It was impossible to tell at first sight whether they were British Pakistanis or locals.

In the 1960s thousands of Mirpuris left to work in the factories and mills of cities like Bradford and Birmingham. Some estimate that around 70 per cent of all British Pakistanis can trace their origins to the region. Practically every household there has relatives in England, I was told. So, when cousins visit from the UK, young Mirpuris copy their style.

My guide was a local YouTube star, Arsalan Shabbir, who has built an online following in both countries. Aside from his crisp white *shalwar kameez*, Pakistan's traditional long tunic and baggy trousers, he looks every inch the modern urban Brit. The sides of his hair are completely shaved; he has his initials tattooed on the back of his neck, a fake diamond earring and a cheeky smile accentuated by a gold tooth. But when he opens his mouth he speaks a thick Pakistani dialect which I can barely follow. He is most certainly a local lad.

We follow him to what he says is Mirpur's premier barbers. A stylish young man is having his designer stubble plucked with tweezers, hair by hair, so the edges of his beard form a perfectly straight line. Both he and the barber also have skin-fade haircuts. Raja, the man being plucked, tells me the style may have originated in Britain, but it's now become identifiably Mirpuri, instantly recognisable elsewhere in the country. He has never been to England himself, but has relatives in Birmingham. 'Do I look like a Pakistani?' he asks me, as he flicks through a gallery of selfies on his phone. 'Could you tell me apart from any British kid?' He's clearly proud to look like one.

But young men here have a complex relationship with Britain. One customer tells me how he was fleeced out of a small fortune by a girl in Leeds who claimed she would marry him in order for him to get a visa. Like many in Mirpur who see their relatives building large homes or businesses with their savings from Britain, he hoped he could find work in the UK. But she took his money and disappeared. Another in the salon is due to get married in a few months – to his cousin in England. Many in the British Mirpuri community still marry within their extended families and some still prefer a spouse from 'back home'.

Some of Arsalan Shabir's videos criticise British Pakistanis for the way they talk about their Pakistani relatives. They certainly use some less than respectful slang. A 'freshie' means 'fresh off the boat' – that is, a recent immigrant who doesn't know anything about life in Britain. A 'mangy' is short for a *mangetar* or a fiancé – in other words, someone who only got into the UK via an arranged marriage.

Arsalan says he gets on well with most British Pakistanis, and as he showed me round his home town we were stopped by three fans who turned out to be from Bradford. Sauntering around with sunglasses and cigarettes and posing for pictures, they looked like they could be on a lads' holiday in Magaluf, except two of them were in *shalwar kameez*.

Arsalan clearly loves the attention – and perhaps the validation – from these British fans. But he says he also resents those British Pakistanis who look down on their freshie or mangy cousins. 'Where do they think their parents come from?' he says indignantly. 'The same place as us.'

Back at the barbers and it's my turn. As he brandishes the electric razor, Irfan the barber tells me that he was the first in Mirpur to start offering skin fades. 'The other day,' he boasts, 'a British customer was so impressed he offered me a job ... in London.'

India

Soutik Biswas – Dalits – 06/08/10

Tensions between different castes are nothing new in India – and neither is caste-based resentment. Increasingly, the Dalit or 'untouchable' families at the bottom of the social pyramid are speaking out and demanding better access to education, better jobs and representation in government. Yet even as India's economy modernised and younger Indians became more globalised in their outlook, at the local level old privileges and priorities were often jealously guarded. When they were trespassed upon, violence could erupt with extraordinary speed and venom. Soutik Biswas heard how easily petty grudges could escalate.

On rainy days, a house often looks gloomier than usual. So it is with Mehul Vinodbhai Kabira's two-room, low-slung home in Gujarat. The house sits on a lane in Bhayla village, a nondescript hamlet where both Dalit and upper-caste people live. It hasn't had a lick of paint in years. The walls are damp. Flies have invaded the place like noisy drones. Under a moss-covered tiled roof are a veranda and some living space.

A rope bed and a blue plastic water tank are the only belongings to be seen. The second room doubles up as both kitchen and bedroom. It is also stacked with the meagre assets of an untouchable. Standing on concrete counters are rows of metal utensils, steel and copper, plates and containers – mostly gifts from family weddings and ceremonies. Old, fraying clothes hang from wall hooks.

In the corner, there's a grey refrigerator, the family's most expensive possession. I open it, to find out what they eat. Inside, I find sixteen bottles of cold water, four pieces of rotting apple and a small pouch of milk. Life for untouchables may have improved over the years, but not enough to afford a fridge full of food.

A small, dusty TV is squeezed into an alcove. Next to it there's a big speaker, and an Indian-made music player. 'These days we buy our music on a pen drive and plug into the player. Then we play it loud and dance!' says one of Mehul's young relatives. But the upper-caste people in the village don't much like the idea of the untouchables enjoying themselves.

Mehul's parents, struggle and deprivation etched on their faces, have worked all their lives as scavengers, picking up manure. Mehul, who is in his twenties, tells me that he decided to shun this life of indignity very early on. So he took out a bank loan and bought himself a three-wheeled auto rickshaw to use as a taxi. He took it to the village bus stand on the highway to try and find customers. There, he says, he was threatened by upper-caste drivers from his village who told him to leave. When Mehul stood his ground, the men came to his home and threatened to burn it down.

'They could not tolerate an untouchable plying his trade in the same place as them,' Mehul tells me wryly. But he didn't buckle. Instead, he left the village to live some distance away, and carried on driving his three-wheeler for four more years.

Two years ago, he returned to Bhayla village. 'I didn't take any chances,' he says. He didn't want any more trouble. So he sold off his three-wheeler, repaid the loan and signed up for work in the housekeeping department of a pharmaceutical factory on the highway. In other words, a cleaning job.

Not so long ago, Dalits like Mehul would suffer unrelenting, lifelong violence and discrimination from other castes. But more recently, affirmative action and political empowerment have improved their status and given them a modicum of dignity. Optimists point to a clutch of home-grown Dalit entrepreneurs who have succeeded against all odds. (Of course, just being upper-caste does not always mean that someone is rich.)

But Mehul's story also demonstrates how the life of an untouchable in modern India is a story of two steps forward and one step back. Here is a man who unshackled himself from the clutches of his lowly, caste-based occupation and became the master of his own destiny. Today, intimidated by his caste superiors, he has slipped back to the status of a contract worker, cleaning factory floors, for a minimum daily wage.

Mehul's near neighbour is a wiry Dalit farmer called Dayabhai Kanabhai Kabira, who told me his story over some sweet tea. He is uneducated, but had the good fortune to inherit some land from his father, which he managed to sell and then buy more land elsewhere to cultivate. As his income grew, he built a second storey on his tiny plot in the village to house his extended family.

What happened next, he says, was shocking. His next-door neighbour, an upper-caste farmer who lives in a shiny multi-storey building, built a six-metre-high wall between the two houses. It was a brutally obvious symbol of upper-caste contempt. 'He built the wall, just so they didn't have to see my face any more,' says Dayabhai. Like many people in Bhayla – and in India – the neighbour simply could not accept the idea of a Dalit who was upwardly mobile.

Rahul Tandon – Faking It for Pushy Parents – 09/08/12

India had always been a status-conscious society – but as consumerism took hold, the status symbols you displayed became ever more important. Perhaps the most prestigious of all, in a country obsessed by academic success and school league tables, was a good education. Sharp-elbowed parents were jockeying furiously to get their children into the best schools. As an expatriate in India, Rahul Tandon hadn't appreciated how intense the competition was, until his turn came to try and get his own child admitted to an elite academy.

I have been living in Calcutta for almost five years. I am a British Indian – born in Britain but with parents originally from India. For most of my life I have been a proud example of someone who has failed the so-called 'Tebbit test': who do you cheer for at cricket? (For me, it's always been India.) So when the opportunity to work in India arose, I jumped at it.

My nine-year-old son goes to an international school that

follows the British syllabus. It's excellent – and expensive. When the time came for my five-year-old daughter to start her education, we decided to give the Indian system a go. My wife joked that it would be a good social experiment.

When we told our friends of the plans there were lots of raised eyebrows. One of them said to me, 'Are you sure – I mean, are you feeling OK?', while another just sat me down and ordered me a very large whisky.

It is not that Indian schools are bad. A friend of mine who put his son in the Indian system told me that when he returned home to the UK, his son was way ahead of the other kids. It is just that getting your child into a good one is a challenge – or, as a teacher friend of mine says, a miracle. Most of the better schools here are a legacy of the British, set up by missionaries. If you can get your child into one, parents here truly believe, they'll have a head start in life.

How hard can it be, I thought? And so I began my quest to find out.

I went to meet some friends who had just got their son into one of Calcutta's best boys' schools. I asked them how they had done it. The father thought deeply for a minute and then said, 'There are two ways: you can pay a donation to the school' – a bribe – 'or you can try and impress them.'

My heart sank. I could not pay a bribe – that would be wrong. But out of interest, I asked him as casually as I could how much that might be. 'Well,' he replied, 'the rumour is that a seat at one of the city's top schools was recently sold to the highest bidder for over £17,000.' I nearly fell off my chair.

'So the only choice that you have,' my friend said, 'is to impress them.' My heart sank again. I had hoped that any need to impress people was over the day I had persuaded my wife to marry me. 'Don't worry!' my friend said. 'There are training centres for parents like you.' Without wasting any time, he gave me a number. 'Call them.'

I did and an appointment was fixed. When I arrived I was met by a smartly dressed middle-aged Indian woman.

Nervously, I blurted out, 'What do I need to do?'

'Well,' she said sternly, 'for a start you should wear branded clothes.' She was clearly not impressed with my sartorial choices. Before I could take offence, she added, 'And your wife' – she paused – 'you are married, aren't you?'

'Yes,' I replied.

She breathed a huge sigh of relief. 'Well, she should wear some jewellery, not anything too gaudy – something classy, maybe diamonds . . . that always goes down well.'

I bet it does, I thought.

'You see, what you wear should portray your financial stability and your class,' she continued. 'You need to stand out from the other parents. Maybe you could carry the latest phone and an expensive pen?'

I looked at my battered mobile and my biro and thought: I am doomed.

It was then that I decided to pluck up some courage and said, 'Surely it's about my daughter, not us?'

She smiled with only the slightest hint of condescension as she cooed, 'That is not how it works here.'

She wished me good luck and said, 'At least your English is good!' as she rushed off to her next client.

I left disheartened. As so often in India, in the end it came down to money rather than ability. Did my daughter get her admission? Yes, she did. Did we follow the advice? No, we did not. But she got in anyway, because we are foreigners and that was considered good for the school's status. As one of my friends said later, 'You're lucky you're *not* Indian – or you would have had to perform like you were in a circus, like we do.'

Rupa Jha – Rape Outcry and Silence at Home – 14/09/13

One of the most visible new currents in reporting about India in the twenty-first century has been the increased attention to women's status – and the growing outrage over sexual abuse and assault. In December

2012, a twenty-three-year-old student was gang-raped, tortured and murdered by six men travelling through southern Delhi in the same private bus as she was. The horrific details of this attack were relayed across the country, and public anger boiled over. Even far from the capital, huge crowds of demonstrators turned out to denounce sexual violence and a justice system which seemed slow to recognise and punish it. Women and girls, newly emboldened, were encouraged to speak out rather than stay silent. Since then, the #MeToo movement has gained ground, with millions more women feeling able to discuss sexual harassment in public – whether it happened at work, in the street or even on a Bollywood film set. But Rupa Jha wondered whether the noisy outcry on the streets was missing the real roots of the problem – which all too often lay at home.

I don't have many particularly vivid memories of my early childhood but the images of sexual abuse never go away. I was barely seven years old, living in a typically large middle-class family with five siblings and numerous other relatives sharing a two-bedroom flat with a terrace. My mother, a housewife, was always busy running such a big family with very limited resources. My father had so much to do just to feed all of us.

I loved being a part of this huge family. It gave me a strong sense of 'home' – but it also made me easy prey for sexual abuse. Distant relatives and cousins kept coming and going through the family home. Being the youngest girl in the family, I was 'loved' by them. These 'love' sessions would happen only when I was alone with one of them. I hated it but, like many others in the same situation, I was too petrified to talk about it. Getting rubbed, touched or kissed, or being locked in bathrooms – that was the 'love'.

Even though the house was always full, I felt completely lonely and violated. I remember the feeling very clearly, along with the moment when it all came to a head for me. I must have been around ten years old. After years of these regular assaults, one day I just broke down and started howling, sitting on the floor. The offender was shocked and petrified. Why was I making a fuss after so long?

What's changed? I remember my elder sister hitting him, hard, until she tired herself out with the effort.

My parents immediately asked him to leave the family home. And that was just about it. End of chapter. After that we all carried on as if nothing had happened. I am not sure what impact it had on me. I didn't start hating men. I grew up a happy child. I became independent. My parents must have felt that all was well with me and the world.

Later, I began to wonder why they never reported the matter to the police. Why was he allowed to go scot-free? Was my otherwise liberal, modern family worried about honour? 'Family honour' – the rationale often used to silence a woman or girl – had never been brought up, but it seems in my case it became more important than facing up to what happened to me. Or was it really about something else?

Decades after the trauma, I have not yet asked my parents these questions. I still don't know the answers. But talking to my sisters, cousins and friends, I discovered a sorority of the abused. So many of them suffered similarly harrowing experiences – experiences of abuse which were followed swiftly by experiences of silence, forgetting or pretending these things did not happen at all.

Then the news came in about the four of the guilty men here being sentenced to death for the fatal gang rape of a student in Delhi last December. I wondered again: when will the code of silence about abuse in Indian homes be broken?

The Nobel laureate economist Amartya Sen famously talked about 'the argumentative Indian'. In a country of a million mutinies and conversations, this is very true. But, as one commentator recently said, we seem to be argumentative about everything except our own selves. When will our families begin debating the roots and causes of sexual abuse – of young girls, and boys, and women?

There's now a conversation happening about how to make women safer in public spaces. This is heartening and encouraging. But while the case of the twenty-three-year-old student was a random attack on a stranger, Indian police records state that in more than 90 per cent of cases of sexual violence the offenders are known to their victims. So what is to be done?

Clearly homes and neighbourhoods are more unsafe than public spaces. That was certainly my own experience. So this is where the change has to happen. India needs to talk to itself more about making its homes safer for women. And to make sure that my six-year-old daughter's generation will not have to live with unanswered questions and silence.

Mark Tully – Modi's Landslide – 22/05/14

Indian politics is garlanded with superlatives: the world's most popu-lous democracy, the world's largest electorate, the most complex elections – and in the case of the Bharatiya Janata Party, the biggest political party anywhere. In 2014, the BJP had been in opposition for a decade, and was spoiling for an electoral fight. This was no insurgent or out-sider movement, but a party with established form: it had been part of national coalition governments before, in the 1970s and 1990s, and had a long history of formidable grassroots organising. But its fiery Hindu nationalism, not to mention its leaders' aggressive rhetoric against Muslims, meant its image abroad was tainted. That turned out not to matter at home: the party's campaign consolidated its support among all social classes in the north of the country, producing a dramatic victory for its leader, Narendra Modi. The BJP won by a landslide, with the first outright majority any party had earned itself in the Lok Sabha. Mark Tully was born in India, and has reported for the BBC from there since 1965 – so had plenty of insight into how the BJP and its leader had taken power. As he explained, the complacency of others in the field had also played its part ...

One of the Congress Party's many blunders during the election campaign was committed by a Cambridge-educated MP who scorned the idea of 'a tea-seller governing India'. The tea-seller in question was the sixty-four-year-old Narendra Modi, who has risen from the obscurity of his job in a teashop in the western state of

Gujarat to rule India. Modi's campaigners immediately arranged for him to speak by multiple video-links to tea-drinkers in shops around the country. This was just one of many brilliant technological innovations the BJP used during the campaign.

The most expensive publicity campaign in India's electoral history, spread across all media, projected Modi as the man who would deliver rapid economic development and called for the election of a Modi, not a BJP, government. One frustrated local BJP organiser told me his candidate, a *sadhu* or Hindu holy man, was refusing to campaign at all, claiming that wasn't what men of his kind did, and that anyhow it wasn't his election, it was Modi's. Modi did indeed sweep the candidate to victory.

Modi has been chief minister of his state of Gujarat for twelve years. Although an investigation ordered by the Supreme Court has cleared him of acquiescing in savage anti-Muslim riots which occurred during his regime, Congress has portrayed Modi as the man who would break India by undermining its secular tradition. One senior member of the party who disapproved of this tactic described the threat to secularism as a 'tired old issue'. Tired and old it proved when compared with the bright future Modi promised.

Also tired and old: Congress's traditional reliance on the Nehru-Gandhi family, when they put up Rahul (the son of party president Sonia Gandhi, and grandson of the legendary Indira Gandhi) to oppose Modi. Even in his own constituency, the response to Rahul was so poor that his charismatic sister Priyanka, who has Indira Gandhi's striking appearance, had to take centre stage. She pushed Rahul off the front pages of the papers, until he was restored by a picture of both of them leaning out of a vehicle – covered with flower petals, waving and smiling happily, looking more like a honeymooning couple than two politicians threatened with disaster.

The Congress Party seemed to have a death wish. Senior members were unwilling to take part in the contest. The finance minister refused outright. In Lucknow, the capital of Uttar Pradesh, I asked a Congress spokesman how the campaign was going, expecting him to reply with at least some sign of optimism. He replied, 'Oh, we are not depressed. We have been badly defeated before, but we have always come back in the end.'

There's no sign of fresh thinking in Congress: the party has already refused to accept the resignation of Sonia and Rahul Gandhi, so there isn't going to be any change there. But if the Lucknow spokesman's forecast is to be fulfilled, the Gandhis will have to relax their control over the party and allow new, strong leaders to emerge in the states.

Modi's promise of development and change has raised expectations so high that he will need to show results quickly or disappointment will set in. Corruption is one of the issues that did for Congress. In tackling that, Modi will be handicapped by the corruption within his own party. Nearly one third of the new BJP MPs have criminal charges against them. A leading party member told me he was all set to be a candidate himself . . . until someone else paid the equivalent of more than half a million pounds for the constituency.

With Modi's background as a former worker of the Hindu nationalist organisation known as the RSS, and his party's links with that organisation, there are inevitable fears that if he does disappoint on other matters, he will fall back on its historic agenda. That will undermine the secular tradition which maintains harmony between Hindus and Muslims, who account for 15 per cent of India's population.

Hinduism hasn't been entirely absent from this election. I have seen a text message sent out on one of the voting days telling Hindus to come out and vote because there were long lines of Muslims at the polling booths. But many have argued that the institutions of India are strong enough to protect the secular tradition, and in his victory speeches Modi has stressed that he wants to take all Indians along with him.

Jill McGivering – With the Cow Vigilantes – 10/12/16

As the BJP government brandished its Hindu credentials, the welfare of cows became a life-or-death issue. Concerns grew about groups of right-wing Hindu vigilantes, who were accused of vicious attacks – mostly on Muslims, but also on critical academics or dissident writers. Any perceived slight to Hindu deities or sensibilities could become a death sentence, if a large enough mob could be stirred to action. In Hinduism, cattle are revered, since 'mother cow' feeds and helps so many millions. Since only Muslims or untouchables – already vulnerable communities – would deal with the remains of a cattle carcass, they were often implicated in alleged illegal slaughter. Jill McGivering went to northern India to investigate what drove the vigilantes.

Mr Tiwari is a lean man in his late sixties with an erect military bearing, metal-rimmed spectacles and a hunter's camouflage cap. As we drove through the north Indian state of Madhya Pradesh, he dominated the vehicle from the front passenger seat. A cotton scarf wrapped round his head also concealed most of his face.

Behind me, right in the back, a thick-set man with bulging muscles and alcohol on his breath was shouting urgently into a mobile phone. He seemed pumped up with adrenalin as well as drink; the mood in the car was volatile. Only Mr Tiwari, directing his team, listening to the shouts from the back and occasionally intervening, seemed in control.

It was late at night and I was out on a so-called cow protection patrol. Mr Tiwari and his men are all Hindus, passionate about the sanctity of the cow. Cows are sacred in Hinduism, and in many Indian states, including this one, it's illegal to slaughter a cow – and illegal to transport one to be killed elsewhere. These men saw themselves, they told me, as protecting not only the cows but their Hindu way of life too.

Mr Tiwari says he runs a network of about 1,300 men in this area who pass on information about the movements of suspected cow smugglers and regularly go out on patrol to track them down. The frantic phone calls in the car were about a recent sighting. Suddenly, a cry went up from the back as the youngest in the team, a mild-mannered twenty-two-year-old, identified a passing truck as their target.

We swerved in the narrow road and turned to give chase. A series of metallic clinks rang out from the front seat as Mr Tiwari loaded his semi-automatic rifle.

The truck in front veered off the main road and careered through the countryside, bouncing and swaying as it sped along the bumpy dirt tracks. Mr Tiwari pointed excitedly. 'Look! Look! You see?'

Horns and the dark silhouettes of cattle were just visible in the darkness. A few moments later, someone unseen in the back of the truck pulled down a thick cover and the cows vanished from sight.

It was a hair-raising chase: the vehicle seemed to know every twist and turn in the unlit landscape. Once, the truck careered straight across a highway, with several lanes of oncoming traffic, and disappeared across the ditch on the far side. Even worse, we followed.

Eventually, Mr Tiwari announced that the men in the truck seemed to be heading to a cattle market, where the smugglers might get back-up. A confrontation seemed likely. The chase was called off.

Afterwards, the vigilantes gathered at the roadside and their tension dissolved into laughter and, for some, more drinking. They passed the rifle from one to another. Why were they taking part in these raids, I asked? 'I am a Hindu,' one told me. 'Mother cow is divine. We won't allow its slaughter in any circumstances.'

When we were alone, Mr Tiwari told me more. Most of the smugglers were Muslims, he said, and killed cows just to make money. This state, Madhya Pradesh, was a major transit route for them. The police didn't stop them, he added bitterly – so his men had to take the law into their own hands.

Cow protection squads have mushroomed in India in the last two years and they're highly controversial. Armed Hindu men, claiming they want to protect cows, have been involved in some shocking incidents: beatings and murders, often targeting members of the minority Muslim community. Sometimes the victims were herding or transporting cows. Others were attacked on trains or in their own homes, wrongly accused of eating or possessing beef.

The mob violence was eventually condemned by the prime minister, Narendra Modi, a few months ago. That was important because his right-wing government is generally seen as pursuing a pro-Hindu agenda and critics say it inspires the groups. All this is particularly sensitive in a country which has seen centuries of tension and division between Hindus and Muslims. Tensions that occasionally, as I've witnessed myself in the past, erupt into bloody communal riots.

Muslims and Christians I met in Madhya Pradesh were certainly frightened. They said attacks on their communities had increased over the last two years – sometimes physical assaults, sometimes legal charges which they said were bogus. They told me Hindu extremist groups were emboldened, growing in strength and able to act with impunity.

Back in Madhya Pradesh, Mr Tiwari and his band of Hindu brothers said they were determined to continue their night patrols, despite the prime minister's warning. 'Day or night, whatever the danger, we will do our duty,' one told me. 'We will fight.'

Megha Mohan – Temples and Pads – 10/01/18

Another story where Hindu tradition seemed at odds with modern India: just who was entitled to worship in a temple? Even for deeply devout Hindus of high caste, there could be a disqualifying factor. Menstruating women were still barred from many sacred sites – and in some cases, all women of menstruating age. In 2019 huge crowds of angry protesters converged on the Sabarimala temple in Kerala state, southern India, as women tried to break a centuries-old ban and go in to pray there. Although the country's supreme court declared that 'to treat women as children of a lesser god is to wink at constitutional morality', for many believers this was a step too far. At another of India's most famous religious sites, Megha Mohan faced up to similar squeamishness.

'Does anyone have a tampon?' I asked as I left the bathroom.

Several of my relatives, who had been chatting animatedly over mugs of hot sweet tea, abruptly fell silent. We were all packed into a modest hotel room in Rameswaram, an island off the South Indian state of Tamil Nadu.

My aunt stood up to reach for her handbag, pulled out a sanitary towel and handed it over to me. 'This will tide you over until we can stop by a pharmacy,' she said, and looking rather sadly at me, she added, 'But you know what this means, don't you? You won't be able to come to the temple.'

Our family were not in Rameswaram on holiday. Although tourism and fishery are the two dominant industries on the island, we were there for a very specific and sombre reason. It had been a year since the death of my beloved grandmother, the glue keeping our intercontinental family together, even though we live all over the world.

My family had observed all the South Indian Hindu rituals: we'd brought her body back to our family home, wrapped her in white cotton, laid her on a large banana leaf and prayed together. The men took her body away for the cremation – a moment I wished I could have attended. But women . . . can't be there.

We didn't eat meat for fifteen days after her death. We performed a special ceremony ninety days afterwards. Then we said our goodbyes at the airport, promising that we would meet again for the final stage of our mourning, at Rameswaram – a well-known Hindu pilgrimage site, famous for its historic temple, perched on the Bay of Bengal.

During the three flights and the bumpy three-hour car journey to get there, I'd read about the temple's long ornate corridors, sacred towers and huge engraved pillars. So, suddenly, snapped back into the moment in my aunt's hotel room, holding her sanitary pad, having her tell me that I was now again excluded from the final part of our shared family mourning, I found myself feeling defensive.

'You're saying I can't come to the temple because I'm on my period?' I asked curtly. She narrowed her eyes swiftly enough for me to know that that tone hadn't been welcome in my youth, and most certainly wouldn't be tolerated in adulthood either.

'I'm sorry,' I conceded quickly. 'But you can't be saying that I've flown all the way here to not be able to go with you inside the temple?'

'It's not what *I'm* saying,' she said. 'That's the way it is.'

The decision was made. I would be waiting with the driver outside the temple.

Growing up, I'd been excused from family outings to the temple when I was menstruating. The reasons given varied. One aunt had told me that it was tradition preserved from a time when women didn't have modern sanitary materials. My mother told me it was because that was the one time of the month time when women could rest completely.

Speaking to academics later on, I heard other contradictory explanations as to why women were excluded. One pointed to the practice of *chhaupadi*, where menstruating women are considered impure and unlucky. One priest said that it was actually because a woman was at her purest during this point in her cycle and should not be tainted by being exposed to others – not even at a temple.

Dr Arvind Sharma, a specialist on the role of women in Hinduism, told me that the contradiction was embedded within the religion itself. 'There are certain times when human beings are considered ritually impure. Women are considered so during periods,' he said. 'That's in Smrata Hinduism, the kind of Hinduism found in the Smrtis, the ancient Hindu texts. However,' he went on, 'in Shakta Hinduism, which is the body of Hinduism that celebrates the feminine, the goddesses, menstruation is considered purifying rather than polluting.'

Menstruation has recently become a hot topic in India, with activists running missions to provide sanitary napkins to rural women and younger social media influencers staging meaningful conversations on the subject online.

All this played on my mind as I waited dutifully outside the temple, watching my family head through the gates.

I reached for my phone to contact my cousin in America, who hadn't been able to make it. After reassuring me that we all find our own ways of dealing with grief, she paused.

'You shouldn't have told them,' she said.

'Have you ever been to the temple on your period?' I asked.

'Most women our age have,' she said casually, adding, 'It's not that big a deal if no one knows.'

Indian Subcontinent

Charles Haviland – Sri Lanka – In the North – 04/08/12

For more than a quarter of a century, from 1983 onwards, Sri Lanka fought a merciless civil war against the Tamil Tigers, a rebel group which claimed to be struggling for an independent homeland for Tamil citizens in the north-east of the island. Emerging from successive waves of violent political unrest in the country during the 1970s and 80s, and well aware of how asymmetrical the conflict was, the Tigers used many tactics other rebel movements rejected. Their leader, Prabhakaran, was elevated to something approaching godlike status in a cult of personality. The group was also notorious for its attacks on civilians, executing prisoners of war, and its repeated use of suicide bombers. It killed high-profile politicians, including the former Indian prime minister Rajiv Gandhi in 1991 and Sri Lankan President Ranasinghe Premadasa in 1993.

By the end of 2009 the civil war had come to an exceptionally bloody end, with the Tigers (along with many Tamil civilians) reduced to sheltering in foxholes in their last, shrinking retreats – where the Sri Lankan army blew them to pieces. The Sri Lankan government had denied journalists access to the battlefield while the killing was going on, but three years later Charles Haviland found it willing to let him into the former conflict zone.

I was with the army as they detonated some leftover munitions near a major battle scene. 'Here, crouch behind these sandbags,' they told me as we stood, in protective clothing, a few hundred metres from the detonation site. 'If shrapnel comes anywhere near, just duck.' I decided to retreat a lot further back. The thud was impressive.

We were in the devastated land which saw the last bitter

fighting of the war. Many of the tall palmyra trees had lost their tops. Half-submerged in the arid ground I saw a single flip-flop and a plastic shoe. Who had worn them? This was where tens of thousands of people cowered – trapped between the Tigers, who conscripted their children and shot those trying to escape, and the army bombardments. Large tracts of land are still mined. There are buildings sliced in half, buses with an end missing. Most poignant are the mundane things. Someone's trousers. A plastic chair, some cushions, a rusting bedstead.

I looked over at a damaged Catholic church; there are both Catholics and Hindus among the local Tamil population. It had been cleared of mines and a Sinhalese soldier took me inside. 'I'm a Roman Catholic myself,' he told me – a minority in the mainly Buddhist military. Sections of the roof had been ripped out, leaving a zigzag pattern of light. The soldier stood by the broken glass of a small shrine to Mary and told me he felt consoled that thirteen of the fourteen Stations of the Cross were undamaged.

Earlier I met a man born and bred in this place. He had lost his brother and his sister as the area was besieged. 'We were climbing over bodies,' he said. Only now, after demining, are families beginning to return to their plots in these villages, many first coming on reconnaissance visits from refugee camps. The surviving buildings are mostly unsafe and will have to be destroyed. And people here have little with which they can rebuild.

Not far away, life seems more normal. A Tamil fisherman wades into the Nanthidakal lagoon and casts his net. There is a jetty and fishing boats here, a low causeway across which trucks trundle into the town of Mullaitivu. There is little evidence left of what happened here in 2009. Huge numbers of Tamils fled from the besieged zone across this lagoon. On its shore the body of the dead Tiger leader, Prabhakaran, was filmed – his scalp covered to conceal the fact that much of it was missing.

There is something which jars. There are many new signs, giving the name of the lagoon and the government's account of what happened here during the conflict. Indeed, a 'war tourist trail' is already springing up for Sri Lankan visitors. But nearly all the signs are in

Sinhala and English only – not in Tamil, the local language and, of course, the language of the people who lost so much.

This is repeated all over the area: at a museum of captured Tiger equipment; at Prabhakaran's bunker, now a tourist spot. At the bunker, there are in fact many Tamil visitors, including school-children from nearby and expatriates from Germany. A Sinhalese soldier, fluent in Tamil, acts as their guide. That is impressive. But there's barely a word of Tamil on the signs.

We ask two members of the military why this is. 'There wasn't anyone who could speak the language – but we'd like to put up Tamil signs soon,' one soldier says. Another, though, tells me he's mentioned it to the higher authorities 'but nobody is interested'. What does this say for reconciliation efforts in this country, whose war sprang partly from the fact that Tamils had said they felt like second-class citizens?

Along with the struggle to rebuild, there is little employment available. By chance we met a former Tiger militant. He said there were no opportunities and no jobs. But he didn't want war again.

The army presence is still massive. Travelling across northern Sri Lanka now is like criss-crossing from garrison to garrison. In fact the number of soldiers on patrol has fallen sharply. They have been asked to reduce their visibility. And many of the army-run cafés and shops which had sprung up have now been closed.

But the government says it still fears a resurgence of the Tamil Tigers. And it is not loosening its grip. In one place an officer told us the army's intelligence network had actually expanded. 'We've been asked to keep a register of what people do and where they go,' he said. 'We even monitor schoolchildren's activities. We know what people eat for their lunch.'

Kieran Cooke – Nepal – Where Have All the Young Men Gone? – 05/12/15

The earthquake of 25 April 2015 devastated Nepal, already one of Asia's poorest nations. More than 9,000 people died and villages across the country's famously rugged terrain were cut off for weeks or months at a time. Many historic temples and shrines were reduced to rubble. When the aftershocks came in May, hundreds more people were killed as they were buried in the wreckage or swept away in flooding. But the country's disasters weren't only natural: there were political and economic ills as well. Long-term development, never mind post-disaster reconstruction, had been far out of reach for many Nepalis for years. After the quakes, Kieran Cooke heard how younger people in particular often saw little option but to leave.

Sabin has a dilemma. We are sitting outside the tea shop in his village, chickens pecking around our feet.

'I am supposed to go back to my job in Qatar, building World Cup football stadiums,' he says. 'If I go, I worry about how my family will manage through the winter cold – but if I stay, where will the money come from to build a new house?'

Sabin's home was destroyed in the earthquake earlier this year. Seven months later, his wife and two-year-old son are still living in a temporary corrugated-iron shelter.

At first glance, Sabin's village in the hills – a three-hour bus ride along a bumpy road from Kathmandu, the capital – has a pastoral, contented air about it. A water buffalo lopes past, a bird swaying like a jockey on its back. An elderly man with a young child on his shoulders walks through the bright-green stalks of the rice paddy. In the distance, a wind is whipping snow off the tops of the Himalayas, leaving white brushstrokes on a cobalt-blue sky.

'It's not only about the money,' says Sabin, a wiry-looking twenty-five-year-old, back from Qatar for the first time in two years. 'So many of the men have gone away there's no one left here now to repair the earthquake damage and build the houses. No one to do the pipes, the electricity – they are all gone.'

More than three and a half million Nepalis – well over 10 per cent of the population of this mountainous, underdeveloped country – have left to work abroad over the past twenty years. Most of them are young men like Sabin. 'Look around: do you see what you *don't* see?' he asks. 'No young men – just the wives and the young and old. The rest are in Malaysia or in Qatar, Abu Dhabi, Dubai – everywhere but here.'

The Nepalis who work abroad have a tough time of it, particularly those who go from the cool air of their mountain villages to labour in the heat of the Gulf states, where summer temperatures can top forty degrees Celsius for days on end.

We are brought a vegetable curry, scooping up rice with a hand. The chickens fight over the scraps I clumsily drop on the floor.

'Many who go to the Gulf find they cannot exist in the conditions there,' says Sabin. 'Some die because it's too hot. But some even die of cold – after hours working in the heat, the labourers take a break. They often go into an air-conditioned place and then fall asleep. And they never wake up. We call it the killing room.'

Remittances from workers overseas are vital for Nepal's economy, making up about 30 per cent of the country's gross national product. Most people still live on less than two dollars a day here. Nepal keeps struggling to fight its way out of poverty, but just when things are looking better, another calamity seems to strike.

A ten-year-long civil war which ended in 2006 has been followed by an extended period of political instability and seriously damaged development efforts. People of Indian origin living in the south of the country claim a new constitution discriminates against them: they've mounted a blockade, stopping imports of fuel, medicines and other vital supplies from India. Many in Kathmandu accuse 'big brother India' of bullying tactics and orchestrating political unrest.

We climb the hill to Sabin's house. All that remains are two walls held in place by precarious-looking wooden buttresses. The rest is rubble. 'People are still nervous,' he grimaces. 'When they hear a loud sound, the children often run out of their shelters for fear of another earthquake. My mother says the gods and the demons deep down in the earth are still fighting.' The inside of the shelter is dark and musty. A large flat-screen TV, brought back from far away Qatar, is propped, unconnected to a power supply, against a wall: goats wander in and out.

Days later, I'm leaving Kathmandu's small airport. Under a sign saying 'Migrant workers' there's a long queue of young men, documents in hand, slowly shuffling towards passport control on their way to jobs abroad.

It's easy to spot the new recruits: they're the ones in the smart clothes, looking more like teenagers going on holiday than labourers bound for building sites in the Gulf. On average, more than one and a half thousand like them leave every day.

I look for Sabin among them. Perhaps he's decided to stay in his village after all. But he'll probably go – to sweat in the heat of the Gulf as he builds football stadiums, and to dream of rebuilding his own house in the cool air of his mountain village back home.

Andrew North – Bangladesh – Rana Plaza – 03/05/13

In the Developed World there's been a backlash against 'fast fashion'. Once, a T-shirt costing a pound might have looked like an astonishing deal, but now it's more likely to trigger a debate about the costs and paradoxes of world trade. Surely we can't keep churning out so many low-cost clothes without inflicting a greater cost on the environment? And what about the workers who sew these garments? The problem is complex: in Bangladesh, textile factories have been a crucial source of income for millions of urban women, and brought many of them previously undreamed-of independence. But employers' practices can be grossly exploitative, and working conditions extremely unsafe. That was brought home by the tragedy of the Rana Plaza collapse on 24 April 2013, when an eight-storey commercial building complex crammed with businesses and workshops collapsed. More than 1,100 people were killed and nearly 2,500 more rescued from the rubble. Andrew North saw the devastation one of the world's worst-ever industrial disasters had left behind.

I looked at the heaped and tangled rubble and saw a foot poking out, dark and oily after several days of decay. Moving closer, I realised that what had appeared to be metal posts were more limbs – and they were everywhere, in a grotesque display of the carnage left by this disaster. A floor above, two outstretched arms were silhouetted in a hole, the hands lifted in a gesture of abject surrender.

And hidden behind the concrete slab right in front of me was the body of a woman, her face frozen in an expression of terrified confusion, and her sari drooping down around her . . . because she was hanging upside down.

I tried to imagine what went through the minds of these people – most of them probably garment workers – as the Rana Plaza factory complex suddenly hit them like a giant wave. But this was a fate worse than drowning – because many of them could have been alive for a long time afterwards, trapped by wreckage and unable to help themselves. Many survivors had only been freed after having their arms or legs amputated because the debris held them fast.

I've reported on the aftermath of several earthquakes, so in some ways these scenes were familiar. Yet this has been a different kind of earthquake – an earthquake caused by humans. By shoddy building work. By corrupt inspectors and local officials who turned a blind eye. By factory bosses who forced workers back into the building by threatening not to pay them. But ultimately by the world's appetite for cheap clothes.

The British and other Western retailers who use these factories say they have tight codes of conduct. But enforcing them is difficult. The Bangladeshi government itself admits that it doesn't have enough inspectors to check fire safety. And as this most recent disaster has made clear, no one was checking how these factories were built.

In the garment boom of the last decade, there's been a rush to boost capacity, with many owners cutting corners for a quick profit. When you drive to the clothing production areas outside Dhaka, the angular factory blocks jut from the skyline above the one-room houses where the workers live – and protruding from the roofs you usually see clusters of steel reinforcing rods, ready and waiting for the next level. Rana Plaza was already two storeys higher than it was supposed to be, but it had those same clusters on its roof, waiting for another floor.

There was barely any garment export industry here twenty years ago. Now it's the world's second biggest after China's. Bangladesh achieved that by producing cheaper than China, and Western retailers came flocking. But there *has* been a string of disasters along the way, from factory fires to other building collapses.

Some say it's a shocking indictment, not just of the industry here, but of globalisation in general. Bangladesh's garment trade is arguably the ultimate globalised industry. Nearly all the inputs are imported – from the raw cloth to the sewing machines – while nearly all of the end product is exported. The only thing Bangladesh provides is plentiful cheap labour.

And yet this is an industry that now provides at least three million jobs in a country that badly needs them. It's helped to empower women, who make up most of the garment-making workforce. They are frequently the main breadwinners for their families.

What many fear now is that this disaster will prompt Western retailers to start moving production elsewhere. Some are already considering it. But that could spell disaster for tens of thousands of poor Bangladeshis.

A few weeks ago, we were outside the ruins of the Tazreen factory, where at least 112 garment workers died in a fire last November – the industry's worst accident until now. The victims of that fire had been making clothes for Walmart, among others. The death toll was made worse by factory guards declaring the fire a false alarm and ordering people back to work, then locking the doors.

Most of the survivors are still living nearby, in the shadow of the factory that killed their relatives and friends. And many of them told us they just wanted that factory to reopen.

Sanjoy Majumder – Bangladesh – Blogger Killings – 19/05/16

The modern history of Bangladesh has often been tumultuous – after all, this is a country born in war, after its secession from Pakistan in 1971. Its domestic politics are rancorous, with paralysing, days-long general strikes called 'hartals' threatening one government after another, and a long-running feud between two rival female party leaders, Sheikh Hasina and Khalida Zia, which was nicknamed the 'Battle of the Begums'. There have been periods of martial law and emergency rule, mass election boycotts, and allegations of corruption. What there wasn't, until 2005, was great concern about Islamist violence. All that ended as the organisation called the Jamaat ul Mujahideen or JMP emerged, threatening – and carrying out – attacks on judges, members of the armed forces, foreigners and non-Muslims. By 2016, the people on its hit list included Bangladeshis who simply weren't deemed religious enough.

Kalabagan is a peaceful, well-to-do neighbourhood in central Dhaka. There's a serenity in the air when I visit. A few boys play a game of cricket at one end and the only sound is the evening call to prayer from a nearby mosque.

I've come to a smart apartment building tucked away in one of the lanes. A guard keeps a discreet watch and the curtains are drawn across a window in a flat on the first floor. This is where Bangladesh's most visible gay activist was hacked to death a few weeks ago.

'About five or six men, saying they were from a courier company, asked for him at the gate,' his older brother Minhaz tells me. 'They even had a parcel which was properly labelled and had the company's logo printed on it,' he adds. The men then burst into Xulhaz's apartment and murdered him with meat cleavers, in front of his mother, who is in her late seventies. 'My mother witnessed it,' Minhaz says, his voice wavering slightly. 'She hasn't spoken about it since.'

Xulhaz Mannan's killing was the latest in a string of recent murders targeting secular bloggers, academics, foreigners and religious minorities. They've all been blamed on Islamist groups, with suspicions that they are linked to the so-called Islamic State and al-Qaeda. Since his death, the killings have continued – a Hindu tailor was killed soon afterwards and a Buddhist monk was murdered last weekend. Like the previous victims, both were hacked to death.

Their deaths have stirred and shocked this nation of over 150 million people. More than 90 per cent of Bangladeshis are Muslim. But its ethnic-Bengali culture was seen as a far stronger national identity than religion. Now there are many in Bangladesh who believe this moderate Muslim state is under threat from a more radical strain of Islam, propagated by groups who are organised, well-funded and inspired by global jihadi organisations like IS. And people like Xulhaz Mannan are easy targets.

'This is a conservative country – not radical but not that open either,' says Brigadier Shakawat Hossain, a retired army officer who's now a security analyst. 'So if radical groups attack people who are seen, in the eyes of some, as un-Islamic because of their religion, orientation or public position – not many will condemn them.'

Perhaps that's why the Bangladesh government has been diffident in its reaction to these murders. There have been very few arrests and the government completely rejects the idea that IS or al-Qaeda have a presence in the country. 'Bangladesh today is safer than Europe or America,' the country's information minister, Hassan ul Haq Inu, says defiantly when I meet him.

But there's no denying that the violence has left many fearful. Imran Sarkar is the country's most famous blogger. Since 2013 he has led campaigns against Islamists and pushed for Bangladesh to become more secular. These days he's been driven underground. It takes me a while to contact him. Meeting him is even more challenging.

After being given directions over the phone, in stages over several hours, I finally track him down to an apartment that is his temporary home. Several security checks later, I am let in. Imran is cheerful but anxious. It's soon clear why.

'I get a death threat almost every day,' he says, scrolling down his Facebook page to show me a torrent of alarming abuse. One message tells him to prepare for his imminent death. 'I got a phone call recently from what looked like a UK number,' he says. 'The caller refused to identify himself and told me I would be killed very soon.'

I ask him if he's frightened and would like to go abroad. He laughs. 'Why would I leave?' he says. 'I need to stay and fight. They've declared a war against my country.'

It's no overstatement. Earlier this month, eight Bangladeshis were deported from Singapore for allegedly being part of a group called the ISB – the Islamic State Bangladesh. The next day, five others from the same group were arrested inside Bangladesh – all of them were said to be plotting to carry out a series of spectacular attacks in the country.

'IS and al-Qaeda may or may not exist in Bangladesh,' says Brigadier Hossain. 'But their ideology has travelled: the pattern of the killings establishes that. And our country has become a competing ground for them – to find fresh recruits and to start a new war.'

Part V:
EAST ASIA

Rohingya Crisis

Jonathan Head – Myanmar – Roots of the Rohingya Conflict – 10/11/12

Decades of military rule left Burma (renamed Myanmar in 2010) to languish on the sidelines as South-East Asia prospered. It stagnated under stifling censorship, inward-looking policies and iron control by the generals in charge. Since it took over in 1962, the ruling junta has fought numerous ethnically based militias and splinter factions, trying to impose control from the centre and exploit longstanding tensions in the periphery. In Rakhine State, divisions between Muslims and Buddhists had grown up over decades, if not centuries, and were particularly bitter. Even as the country appeared to reform itself from late 2010 onward, with the release from custody of long-term prisoner of conscience Aung San Suu Kyi and various elections, this dynamic never improved. After a flare-up of communal violence where dozens of people were killed and many thousands displaced, Jonathan Head looked into the root causes of the conflict.

In early 1949, a former British administrative officer who had served in what was then known as Arakan wrote a confidential memo to a colleague in the Foreign Office. He offered his insights into the unrest which had swept through that region of western Burma after independence the year before.

He described two peoples. He likened the Arakanese Buddhists to the Scots. They were descendants of a powerful Buddhist kingdom in the sixteenth and seventeenth centuries, and had provided some of Burma's ablest politicians and civil servants.

Then he described the Muslim 'Chittagonians', as he called them – people of Bengali origin. 'They are fanatically religious,' he wrote, 'and more hard-working than the Buddhists.' They had been moving steadily south into Burma for centuries.

He went on to explain that when the British had retreated from the Japanese army in 1942, there had been terrible massacres in which thousands had died. The Muslims had by and large backed the British, the Buddhists the Japanese.

After that memo was written, Burma endured more than a decade of civil war and shaky democracy, before the military took over in 1962 and turned inward. The Burmese people were shut away from the outside world, trapped in their poverty and with their historical memories. Burmese history was in some ways frozen.

The Shwe Zedi monastery is a shabby but dignified old building in Sittwe, the provincial capital of Arakan, now known as Rakhine State. It has a great pedigree: it was the home of Burma's first political monk, who inspired Aung San, the father of opposition leader Aung San Suu Kyi, to begin his campaign for independence from the British.

The monastery also played a courageous role in the mass protests against military rule five years ago. Today, though, it is supporting something altogether less heroic – the drive by local Buddhists against Muslim Rohingyas, the 'Chittagonians' of the old British report.

The abbot, Ashin Ariya Vunsa, ran through a long list of grievances for me. 'The Rohingyas take too many wives and have too many children,' he said. 'They don't educate their women, and they violate ours. They want to turn Rakhine into an Islamic state.'

These views are widely shared in Burma. And they are given legitimacy by the government's refusal to recognise Rohingyas as Burmese citizens. Buddhists feel entirely justified in calling for Rohingyas to be expelled. In a country brutalised and impoverished by the long years of military rule, this has escalated all too quickly into something like ethnic cleansing.

I travelled around northern Rakhine the only way you can – by boat. The coast here is indented by a jagged maze of creeks, so shallow we often had to wade a long way to reach the isolated fishing villages. At Sentaymaw, normally a community of around fifteen hundred, we found 5,000 displaced Muslims, some still in shock from watching their neighbourhoods being burnt down. There were men with bullet and machete wounds, and a very young widow with two small children who'd seen her husband shot dead in front of her. There was barely enough food to go round, even with deliveries starting to come in from the UN. Malnutrition is visible in the children.

Further upstream, we saw the scorched palm trees that indicated what had been the Muslim quarter of Pauktaw. Only the walls of the mosque were still standing; the wood and palm-thatch houses had been reduced to piles of crumpled iron and ashes. You could hardly guess that hundreds of families had lived there up until last month.

The one surviving Muslim quarter in Sittwe, Aung Mingala, is sealed off from the rest of the town by barricades guarded by soldiers. Its residents are living just 500 metres from the town centre, but they can't go there, to work, to shop or for medical help. They are living in a ghetto.

Both communities are now afflicted by a paralysing fear of the other side. They won't go near each other, so there's not much chance of easing the mistrust. Government officials say they daren't even discuss citizenship for the Rohingyas for fear of a violent backlash by Buddhists.

Rakhine is the second-poorest state in an already very poor country. Its Buddhist population shares a folk memory of a glorious past, before they were incorporated into the Burmese kingdom, and then the British Empire. They have more recent tales to tell too – however true these may be – of Muslim atrocities, of creeping Islamisation. It's a poisonous cocktail of resentment, pride and prejudice, and it has cast a dark shadow over Burma's new democracy.

Justin Rowlatt – Myanmar / Bangladesh – Farmers and Fleeing People – 02/12/17

For the Rohingya, 2017 was a year of unrelenting horrors. By its end, more than a million of them had fled over the border into Bangladesh, fleeing the violence in Rakhine State. Their persecutors weren't only the Myanmar military: many were civilians, their former neighbours or even friends, who'd stood by as Rohingya homes, businesses and mosques were burned. There were credible accounts of massacres, rapes and torture committed by both troops and civilians. Myanmar's government insisted the military was fighting a vicious Rohingya insurgency, led by a rebel group which had attacked police and army posts and killed dozens of their men. The violence, and the exodus which followed, were not hidden from the world. Whether on a grainy mobile-phone video shot by a refugee or in the high-definition images of an international news crew, a desperate text message or a front-page investigative report in the international press, evidence of terrible crimes and great desperation was everywhere. When Justin Rowlatt went to the border he found thousands more refugees stranded on the beaches of Myanmar, still hoping to leave.

Nobody wanted to take us out onto the Naf, the slow-moving river dividing Bangladesh from Myanmar. It was only when we got formal permission from the Bangladesh border guard that a fisherman finally agreed.

Abul's boat was beautiful, a graceful wooden crescent. The ancient diesel engine was not: it coughed out clouds of black smoke as it sputtered into life.

Even from the Bangladeshi side, I could see the smudge of colour at the top of the beach. As we chugged closer, it resolved into a row of tents. There were black silhouettes, stick figures moving: the last stragglers from the great tide of Rohingya refugees that has poured into Bangladesh.

They must have seen the boat coming, because they began to gather by the shoreline. That's when Abul announced we had crossed into Myanmar and were far too close. I could see the barbed wire fence behind the refugees, and to one side, the tower of the Myanmar border guard post, ominous and white against the vivid green hills beyond.

'If we go any closer the guards will shoot,' said Abul emphatically, and heaved the tiller. The bow swung round; we headed back to Bangladesh.

I looked behind us. The crowd of marooned people was slowly dispersing.

They're stuck because a few weeks ago Bangladesh decided to try and stem the flow of refugees. Abul pointed out the charred wrecks of a couple of fishing boats. 'Burnt by the border guards because they'd been picking up refugees,' he told me with a shrug. 'It is meant as a warning to us all.'

Conditions in the refugee camps are improving – there's food now, and basic sanitation – but the truth is, they are really just a giant prison. There are military checkpoints on all the roads. I watched as Bangladeshi soldiers pulled refugee after refugee from rickshaws and buses. 'We've welcomed them in here, but now they want to invade our country,' explained one of the officers without emotion.

It isn't just the Rohingya who are trapped.

I hiked up one of the small hills to get a sense of the scale of the vast camps. Sitting in the shade was a row of Rohingya men chatting amicably together ... or at least I assumed they were Rohingya. They wore the usual *lunghis* (skirts of brightly coloured cloth) around their waists, dusty flip-flops on their feet.

I asked where they had come from, and to my surprise one of them announced that he was actually a Bangladeshi farmer. 'This is my land,' he told me. He made a sweeping gesture with his arm: 'All of this, everything is mine.'

He explained that, like many Bangladeshis, he had spent time working as a labourer abroad, and saved the money he earned in Saudi Arabia to buy a little farm. 'It wasn't great land, lots of hills,' he told me, 'but I had planted it with orchards and there was bamboo. Now look at it!'

And suddenly he laughed. The Rohingya refugees sitting with him laughed too, and so did I. We all laughed and laughed. The devastation of his land was just so comically total. Every scrap of vegetation had gone; the whole place was now covered with make-shift shacks made of bamboo and plastic sheeting.

'When the refugees first came I tried to fight them off,' he explained. 'I threatened them with my machete, but there were so many of them. Then they offered me rent. They said they would pay me 2,000 taka a month.' That's around £18.

And do they pay, I asked? This produced another laugh. 'They haven't given me so as much as a split lentil,' he said pinching his fingers together. 'And my bamboo, it is all gone; they cut it down to make their houses.'

So why was he sitting with them like this – why wasn't he angry?

'What choice have I got?' he replied. 'They are my neighbours now; we may as well get along.'

The farmer had accepted his fate with stoicism and humour, but lots of other Bangladeshis have not. Take them aside and over a cup of achingly sweet *chai* they'll tell you that the Rohingyas are overrunning the place. They've taken the land, they say, and now they get free food and medical treatment. No wonder so many of them want to come here, they complain. The tension is palpable.

UN officials scratch their heads about how they can make sure that local people benefit too. They know that whatever the governments of Myanmar and Bangladesh say, the refugees – almost a million of them – aren't going anywhere soon.

I managed to call a Rohingya contact who's still on the other side of the river, hiding in Myanmar. He told me the local Buddhists have begun harvesting the fields of their former neighbours. After all the atrocities, this is the final indignity, he said: 'Our crops are being stolen.'

I tried to contact him again a few days ago. He didn't answer his phone.

Nick Beake – Myanmar – It Just Never Happened – 13/02/19

The Rohingya crisis is one of the most clear-cut examples in recent years of a story almost too polarised to tell. Opposing sides don't just see the same facts from different perspectives; they refuse to agree on any common ground at all. Even using the word 'Rohingya' is a provocation to some Burmese, who insist there is no such ethnicity, and that while Myanmar has many 'legitimate' minority groups, the Rohingya are simply illegal immigrants with no rights to be in Myanmar in the first place. Therefore, it follows, there have been no abuses committed against them – no matter what the United Nations might claim. Nick Beake got a dispiriting lesson in the power of denial.

Have you ever headed off on holiday to get away from it all, only to run into someone from back home?

I once bumped into an old university friend in the packed rabbit warren that is Marrakech's main market. George and I hadn't seen each other for five years, but after this chance encounter became housemates back in London. Then there was the moment on San Francisco's Golden Gate Bridge, when I found myself cycling past another friend I'd totally lost contact with. 'Hello, Pete!' I casually called out, before my brain could process the remarkable odds against our meeting.

Recently, I was exploring another tourist favourite, the ancient Burmese city of Bagan: a landscape bejewelled with more than 2,000 Buddhist temples. The reunion here was a little different. It wasn't with someone I knew, but with a local man who instantly took me back to my city of birth in the UK.

It turned out he knew Bristol better than I do. Chatting to him was uncanny: a home from home. Just a few of years living there had been enough for him to become a connoisseur of the best pubs, an authority on the rival football teams, and an aficionado of the street artist Banksy's early works. Amazingly, he'd even got his head round the council's unfathomable recycling system. We discussed all this in perfect English. He even threw in the authentic local greeting – 'All right, my babber' – in perfect Bristolian.

It was bizarre and brilliant having an unexpected pint with this son of a mystical Burmese land who was also so obviously on my wavelength. But then came the moment. The gob-smacking, hold-your-horses, 'did he really say that?' moment.

It's a moment I've experienced time and again in the year I've been in Myanmar. What is it that breaks the common bond? The Rohingya. Whenever I mention them here, I prepare to be unpleasantly surprised.

'All this fake news!' my hitherto kindred spirit despaired. 'I can't believe people fall for it. Those Muslims came over from Bangladesh – most of them are terrorists – and they wanted to take over,' he insisted. 'They burned their own villages down because they knew they would be given more land, elsewhere, in a better place.' Let's just say that's a somewhat different interpretation of events to what UN inspectors found. They believe it was genocide. 'So basically,' my new acquaintance concluded, 'the Burmese army came in to fight the terrorists and then *protected* the area.'

Erudite, educated and in complete denial. This is, of course, the very accusation levelled at the mother of the nation, Aung San Suu Kyi. Myanmar's indefatigable defender of universal human rights is now accused of ignoring ethnic cleansing on her doorstep.

It's startling when you first encounter ministers, civil servants and journalists who just offer a shrug of the shoulders when you mention the alleged genocide. But given time, you expect it. Some local newspapers whose staff battled years of army censorship, dreaming of a free press, will not even mention the Rohingya by name. This is an ethnic group which officially does not exist.

I tried to pin down exactly *why* our Burmese Bristolian thought the Rohingya crisis was made up.

Was it straightforward racism? It's true, Rohingya Muslims are largely detested by the Bamar people, the dominant ethnic group here. For decades, under military rule, the Buddhist majority was fed a poisonous narrative: that these particular Muslims were sub-human. With no counter-argument from the outside world, hatred of the Rohingya flourished. This guy, though, had been given the chance to experience the outside world, to enjoy the diversity of life in a cosmopolitan British city.

I continued: 'They reckon ten thousand Rohingyas were murdered by the military.'

'No, I don't think so,' came his dismissive reply.

'Thousands of women say they were raped ...'

'No. There wasn't enough time to rape all those women.'

By now, any fraternal bonhomie had disappeared – certainly on my part. As I framed my next question, I pictured eleven-year-old

Omar, whom I'd met in a refugee camp in Bangladesh a few months before. The boy who told me he wiped away tears every morning before getting ready for school. Who told me he'd seen his mother, father, three brothers and two sisters shot dead.

'But what do you think about those orphans?' I pressed. 'You know, the ones you see on the news – who describe how their whole families were murdered?'

'No ...' came the reply from my Burmese brother, who had so effortlessly transported me back to where I was from. 'No, it just didn't happen.'

I guess there are some things which are just too close to home.

East Asia

Rachel Harvey – Japan – The Damage – 17/03/11

Covering natural disasters is daunting on every level. Just getting to the scene and organising enough power and connectivity to report is tough. Verifying the most basic facts is an immense challenge: every estimate of casualties or costs can only be provisional, for many days after the event. Staying focused on reporting amid so much human suffering tests many a journalist's energies. And as Rachel Harvey saw for herself in the aftermath of Japan's immense Tohoku earthquake in 2011, the sheer scale of the destruction can be almost impossible to grasp.

It's the big numbers that tend to grab the headlines when a natural disaster strikes. How many people have died, how many are missing, how strong was the earthquake, how high was the wave. But it's the small details that for me bring home the human tragedy and tend to stick in my mind. The family photograph lying in the wreckage of what was once someone's home; a child's shoe; a broken teacup. It's as if the human brain can't quite process the sheer scale of the devastation. So it concentrates on small things, familiar things, that it can understand.

But then there are the images I remember precisely because they don't seem to make sense. There are things about natural disasters that just aren't natural. A white van wedged between two trees, three metres off the ground; a fishing boat marooned halfway up a building; a dead fish on a bathroom floor miles from the sea.

The fish – a sea bass, for anyone interested – was lying next to an orange and a lump of timber in Keizo Shutou's house. Or at least, what was left of his house. The walls and roof were intact, but the windows had gone and every room was filled with debris and mud. And yet this sprightly man in his seventies was determinedly positive. As he picked his way through the rubble, trying to find any belongings he could salvage, he told me he would be moving back home as soon as he could. 'We can clean this up,' he said. 'The structure of the house is fine. We can rebuild it.'

'But,' he added, 'I've never known anything like this before.' Keizo is seventy-four years old. A lot of terrible things have happened in Japan in his lifetime: the bombs which landed on Hiroshima and Nagasaki, the earthquake which destroyed large parts of the city of Kobe, and now this. An earthquake and tsunami that had ravaged Keizo's homeland. And yet he is counting his blessings. 'It's much worse closer to the shoreline,' he told me.

He wasn't exaggerating. Standing on a hillside, I looked into a valley that had once been filled with houses, shops and businesses. I could see the ocean away to my right, the direction from which the tsunami waves had come. And to my left, about as far as I could see inland, there was nothing but wreckage. Part of the hospital and local government offices were still standing. But almost everything in between had been flattened by the force of the torrent of water raging up the valley. Here and there I could see the foundations of buildings lying exposed. In between there were piles of broken timber, pieces of twisted metal, wrecked cars, smashed boats, torn clothes.

I had a horrible feeling of déjà vu. I'd seen this before, in the Indonesian province of Aceh in 2004. I really didn't think I'd ever report the aftermath of another tsunami. But there are some stark differences between what is happening here in Japan and what I witnessed in Indonesia. Aceh had been the scene of a long-running separatist insurgency. There was a huge military presence there before the tsunami, but not much in the way of disaster relief. And the destruction was so extensive, the loss of life so great, it took days for the full scale of the horror to become clear. It took a week before a major international relief operation was established. Indonesia, and the rest of the world, have learned valuable lessons since then.

Japan was as well prepared for a natural disaster as any country could be. It's an industrialised, high-tech, well-educated nation.

People are taught from primary school what to do in the event of an earthquake. There are public address systems and sirens ready to sound the alarm. The number of people who have lost their lives in this disaster is an unimaginable tragedy: the worst to befall this country for a generation. But it's fair to say many more people would have died if those systems hadn't been in place. And I have been struck by the stoicism, patience and organisation of the Japanese people in the face of disaster.

The danger isn't over yet. There are serious concerns about damaged nuclear power plants. Aftershocks still fray the nerves, and petrol and drinking water are in short supply. And yet people are already talking of rebuilding their lives, just as they did immediately after the tsunami in Aceh. Incredible resilience, incredible independence, incredible strength in the face of adversity. That's the image from this natural disaster that I would choose to remember.

Peter Day – Japan – Ageing – 08/03/12

As more and more of the world's developed economies confronted their changing demographics, the challenge of caring for ageing populations became ever more pressing. Policy-makers looked to countries where the proportion of older people was highest for clues on how best to manage the transition. The potential problems it poses are obvious: health care and social services under pressure, employers running out of manpower, dwindling tax revenues. By 2010 a quarter of the Japanese population was over sixty-five years old, and that's predicted to rise to more than a third by 2050. Peter Day saw much to consider in Japan's example.

There is snow in the air up in the hills on Shikoku Island about an hour's flight south of Tokyo. But the chill winds are more than just the weather. Like so much of Japan, Haigyu Village is getting old. Its senior citizens are beginning to feel their age. On a recent Saturday afternoon, a group of them met to talk things over with me, round an inadequate fire in the long, low workshop where a local craftsman still creates exquisite hand-made Japanese paper in sheets rough on the surface and full of random whirls of embedded plants and leaves.

All but one of the six were well into their seventies; the paper-maker was an active eighty-two. Like so many of the ageing people I met in the fields and woods and farmhouses of Shikoku, they seem extraordinarily spry and alert.

Many of them are still busy working to supplement meagre pensions. The women are bent, but disarmingly smooth-skinned and relaxed about getting old. The men are more circumspect, especially those who live on their own, up the steep village lanes. 'Every night I worry about what will happen if I get ill,' said one of them, edging closer to the fire. 'Would my neighbours find me?'

Traditional life in these upcountry villages is under threat. In two places I visited, the over-sixty-fives now outnumber younger people. Haigyu is becoming a ghost town, as the offspring leave for jobs in the city. Buddhist farmers do not like ancestral homes to be occupied by incomers. So their abandoned farmhouses lie empty and decaying in the sleeting wind.

There is much that is wonderful about the extraordinary increases in longevity, especially when the silver-agers remain so fit and active, helped (perhaps) by that Japanese diet of fish and vegetables. But eventually the increasing number of elderly will need care; that will be expensive in a country already more indebted than Greece. By 2050, four out of ten Japanese will be over sixty-five.

And the other side of this story is an extraordinary reluctance to have children. The young Japanese people I asked about this slightly touchy subject were rather vague about the reasons for the baby famine, though many mentioned a feeling of foreboding about the cost of the future.

This doubt comes after two strange decades of slow or negligible economic growth, and weird price deflation following the bursting of the huge Japanese property bubble at the end of the 1980s – to say nothing of the more recent twin catastrophes of March last year.

The shrinking of the Japanese population has already started, and it is gathering pace. By 2050 the total number of Japanese will have been reduced by one third to less than ninety million people;

the young working ones will have a great burden of elderly Japanese on their backs.

Nothing like this has ever happened through natural causes in history. And it's not just Japan. As baby boomers born after the Second World War start hitting sixty-five in many countries this year, Germany, Italy and China will soon join the Japanese as ageing, shrinking countries. Britain is facing a similar trend, but not so sharply, thanks to immigration boosting its workforce. Japan is (to put it mildly) wary of immigrants.

In many countries the next fifty years are going to be hugely different from the decades of economic growth we have enjoyed over the past century. In a shrinking population, growth can no longer be assumed as a default position; in conventional terms, recession becomes the norm. The pressure on prices is falling, not rising. Japan has shown that deflation is devilishly hard to get rid of: why buy things today when they'll be cheaper tomorrow?

Ingenuity and technology may help ease the high price of caring for older people. So I sat on a lavatory that monitors your vital functions on a daily basis; elsewhere there are robot pets to keep ageing minds alive, and remote sensors to alert a son or daughter hundreds of miles away if an old parent fails to get out of bed in the morning. But many Japanese politicians and policy-makers I heard from seemed to be hoping against hope that active old age (and an increase in volunteering) may ease the burden on the state.

Even so, you can't get away from the idea that we are all heading into the unknown. At least that's what the old men told me the other afternoon, huddled round a struggling fire in the paper workshop in Haigyu village.

Rupert Wingfield-Hayes – North and South Korea – Choco Pies Tell All – 04/04/13

When visitors to North Korea have fallen afoul of local laws, the Kim regime hasn't hesitated to expel or jail them. If they come from states which have hostile relations with the Pyongyang regime, the stakes are particularly high. Such detainees have disproportionate value as bargaining chips during international negotiations over nuclear weapons or security policy.

In 2013, a Korean-American tour guide named Kenneth Bae was condemned for allegedly trying to overthrow the North Korean government while visiting the country. (He was released eighteen months later.) His sentencing followed weeks of aggressive comment from North Korea, which threatened both South Korea and the United States with nuclear war. Analysing the sabre-rattling, Rupert Wingfield-Hayes suggested it revealed a gnawing sense of insecurity among Pyongyang's ruling elite.

Not a week goes by without the North Korean regime appearing to commit a fresh outrage. This time it's an American tour guide, sentenced to fifteen years' hard labour. And for what? Last month we were being threatened with nuclear war by a thirty-year-old despot with a bad haircut. It's no wonder the Western media tends to fall back on clichés. 'Mad, bad and sad!' A 'hermit kingdom' peopled by 'brainwashed masses'. I am guilty of using a few myself.

The images we see from Pyongyang tend to reinforce those clichés. Massed ranks of soldiers goose-stepping across Kim Il-Sung Square in perfect unison. Groups of women factory workers weeping at the sight of their young leader. 'Is this adoration real or is it fear?' we ask ourselves. The images build a picture of an isolated, paranoid country, its people kept in ignorance and fear by an all-powerful militarised state.

But here are some other images that have come out of North Korea recently: a policewoman chastising the driver of a souped-up Mini Cooper on a Pyongyang street corner; a businessman (possibly Chinese) driving through the city's downtown in his Porsche Cayenne ... But the most striking image I've seen was a North Korean family in a village near the Chinese border watching a South Korean TV drama on a smuggled DVD. The trade in smuggled South Korean DVDs is now huge. And that can mean only one thing – North Koreans are not nearly as isolated as we tend to think they are.

While I was in Seoul recently, I met with a number of North Korean defectors. What one of them told me left me open-mouthed in amazement. It all relates to Choco Pies. They're round chocolate snacks filled with marshmallow: a bit like a Wagon Wheel, but Korean. And what does this have to do with North Korea?

A few years ago, South Korea set up an industrial park just across the Demilitarised Zone in the North Korean town of Kaesong. The Kaesong factories employ North Korean workers – or they did until last month when Pyongyang shut them down. South Korean companies were banned from paying their North Korean employees cash bonuses. So over the years they started paying them with various food products, including Choco Pies.

But instead of eating them, the employees took their Choco Pies back to Pyongyang and started selling them on the black market . . . for three to four times their original price. North Koreans, it seems, can't get enough of the capitalist confectionery.

Back to our defector. We were speaking in Seoul shortly after the Kaesong industrial park had been shut down.

'I called my dad to tell him there was going to be a shortage of Choco Pies and the price would be going up,' he said. 'I told him he should go to China and buy as many as he could find. That he should make a good profit.'

All of this left me extremely confused.

'I'm sorry,' I said, 'you just said you called your father. Now where is your father?'

'Oh, he's in North Korea,' he said.

'And how did you call him?'

'On his mobile phone,' he replied, as if this were the most normal thing in the world.

'Sorry . . . your father has a mobile phone in North Korea that can receive international calls?'

'Oh yes!' he said. 'It's a Chinese phone. He lives near the border, so he can get on a Chinese network.'

'And how common is this?' I asked.

'Very common. Everyone along the border has them. They need them for doing trade with China.'

'But couldn't he get into trouble for that?'

'Look,' he said, 'there are fifty thousand North Koreans crossing backwards and forwards into China to trade. There are another hundred thousand living in China doing business. What's Pyongyang going to do? It couldn't survive without the trade.'

This is a very different North Korea from the one portrayed in Pyongyang's propaganda films. It's a country where the black economy *is* the real economy, where bribery rather than obedience is now the surest means of survival.

Do North Koreans still believe in the cult of the Kim dynasty? Perhaps many still do. But tens of thousands already know what life is like just across the border in China. Along with smuggled DVDs and Choco Pies, more of them are starting to glimpse what everyday life is like in the South as well.

So I think Pyongyang's recently manufactured crisis does not speak of a regime that is strong and confident, but of one that is weak and scared. Scared of the outside world – and increasingly of its own people too.

South-East Asia

Rebecca Henschke – Indonesia – Crazy Rich Indonesians – 25/10/2018

Indonesia is one of the world's most populous nations, with over 270 million citizens – only India, China and the USA have more. It's also home to a rapidly growing number of people who are rich, or at least comfortably off. After a massive reduction in the poverty rate over the last two decades, one in every five Indonesians now has an income high enough to be called middle class. In the capital, Jakarta, Rebecca Henschke found that she was mixing in increasingly affluent company – and asking more questions about where the money was coming from.

The elaborate, colourful invitation had said it would be a 'dog-themed' birthday party. 'Well, that's cute,' I thought – and different, too. In this Muslim-majority country, dogs are not generally well-liked or looked after. But that wasn't the only surprise. To celebrate their little girl turning six, Eileen's family had turned an entire empty piece of land in Menteng, the most expensive part of Jakarta, into a park for the day.

Security guards escorted us off the street into another world. Real grass – an incredibly rare thing in this concrete jungle – had been rolled out, full-grown trees brought in and an obstacle course for dogs laid out. In one corner, a groomer was giving appreciative canines (also specially brought in for the event) massages and baths. In another was an air-conditioned marquee where the parents were sipping freshly made iced coffees ... The middle of the 'park' was filled with dog-shaped balloons, a bubble-blowing performer and a slime-making station.

I had just got back from reporting the destruction, grief and devastation in Palu, on the island of Sulawesi, which had been hit by a tsunami and earthquake. It made for a surreal, almost bizarre contrast.

'Where do you go from here?' I whispered to one of the other parents. 'What would an *eighteenth* birthday party have to be, if you kept this up?'

'It's not what the children ask for, it's really for the parents,' she replied.

The party bag we left with was three times the size of the present we had brought.

I am not sure why I'm still surprised: parties like this have become the norm amongst the upper-class Indonesian children my kids now go to school with. One family hired a film company to produce a special re-edit of the Hollywood blockbuster *Suicide Squad*, with the birthday girl appearing as a character in key scenes; the kids watched it at a specially reserved showing in a full-size cinema! On that occasion, I had recently returned from a trip to the remote province of Papua to cover a children's health crisis: tiny, malnourished toddlers dying in a measles outbreak.

Many members of Indonesia's growing upper-middle class, concentrated solely in the west of the country, have money their parents would never have dreamed of – and most think it's essential to show it off. They're riding a commodities boom, the burning and churning up of this vast archipelago's rich natural resources: timber, palm oil, coal, gold and copper. This, combined with aggressive domestic spending, low taxes and little enforcement of labour laws, means that those who know how to play the system are raking it in.

Pak Saliman is one of the many who don't understand the 'system' – but has, in a way, also eked out a future for his children that is very different from his own life. He is a street sweeper, paid half the minimum wage of £150 a month to take away the waste of the wealthy houses in Menteng – great plastic mountains in front of Greco-Roman inspired concrete mansions, piles of rubbish like monuments to out-of-control consumerism.

He pulls by hand a cart that he banged together from scavenged wood. He is the strongest man I have ever seen. My children call him 'Superman'. He pulls anything that might have value out of the trash, sorts it and stores it, and then sells it on.

Saliman lives in a room behind our house; he effectively came with the property. He was squatting there at the time we came to look it over before deciding to rent, and asked if he could stay. I am glad we decided, after some debate, that he should; he has become like an uncle to my children. He's a farmer at heart, who has turned our swimming pool into a fish pond and the garden into a banana plantation.

Whatever he earns, he sends to his family in a village in central Java, going home just once a year to see them. That money, from the waste of the rich, has meant that his children finished high school and now have jobs in shops in the giant glittery shopping malls of Jakarta.

'What's an iPad?' he once asked me. 'My son says he really needs one. How does it work?' I talked him out of paying for one, suggesting a cheaper alternative.

His daughter came to stay briefly – she seemed very interested in her phone. She hasn't come to visit since. He is very proud of them.

Jonathan Head – Philippines – The Lady Killer – 27/08/16

Rodrigo Duterte, president of the Philippines since June 2016, is a larger-than-life figure who often leaves the international media at a loss. He's extraordinarily direct – even eccentric – in his rhetoric; his speeches are often larded with dirty jokes, singing, stinging insults, and verbal attacks on revered people and institutions. (In this deeply Catholic country, he once called the pope a 'son of a bitch' for snarling up Manila's traffic during a visit.) Sometimes he's joking, and sometimes his outrageous remarks have been misunderstood, or lost in translation. But when he talks about hunting down and killing drug dealers, he's not kidding. During his years as mayor of Davao City in the late 1990s, he was completely unapologetic about encouraging police to shoot first and ask questions later. Once he was in power in Manila, Duterte went further – in both word and deed – and the bodies started to pile up. Jonathan Head had a deeply disturbing encounter with one of the president's foot soldiers.

There are some wonderful movie assassins; the profession, if you can call it that, lends itself to wicked black humour. But meeting an assassin face-to-face for the first time is a disconcerting experience, especially if she arrives in the form of a lightly built, nervous woman with a baby in her arms.

'I'm sorry,' said Maria – the pseudonym we had agreed to give her – as she arrived at the modest hotel room in Manila where we had agreed she could be interviewed in safety. She didn't have anyone to look after the baby. Fortunately he was too young to understand what his mother was about to tell us: how she kills people for a living.

The Philippines has had a long love affair with guns, with more than 1.5 million legal gun-owners, and many more unlicensed weapons. Law enforcement is weak, and there is a vast and visible gap between the country's wealthy elite and the urban and rural poor. So it should not surprise anyone that contract killers are also a feature of Filipino life – and death.

Before he became president, Rodrigo Duterte was best known for cleaning up crime in Davao, the city he ruled as mayor for twenty-two years, by tolerating, perhaps even encouraging, death squads to target drug dealers. Now he is applying the same formula nationwide. He sounds genuinely angry when he talks about the millions of Filipinos caught up in what he calls a pandemic of addiction to the drug crystal meth. 'If you destroy my country,' he warned the dealers, 'I will kill you.'

That's not to say the new government is actually hiring freelance killers. It isn't that straightforward. What the president has done is to order the Philippines police to be a lot more aggressive with drug suspects. 'If you think they are going to shoot you,' he said, 'don't hesitate to shoot first, and shoot to kill.' But most of the 2,000 street killings since Mr Duterte took office were not carried out by uniformed police officers but by shadowy hitmen (and women) working for equally shadowy bosses.

Maria had never spoken to a stranger about what she does before, and she seemed to find the experience difficult, squeezing her eyes tightly shut and wringing her hands as she struggled to answer my questions.

It was her husband who had got into the killing business when he was helped to get out of jail by a police officer, who she says was also big in the drug trade. The officer wanted people who owed him money killed. But a woman is useful in the contract-killing business because she can get closer to a target without alarming him. So they roped in Maria.

She shot her first victim two years ago. She was very scared then, she recalls, but since President Duterte announced his war on drugs three months ago she has killed five more people, on the orders of the same boss. She doesn't know why they were targeted. As the boss is a big-time dealer himself, it isn't too cynical to suspect he is wiping out other dealers who might incriminate him.

Maria says she doesn't want to consider such questions. 'I just think to myself, I am doing this for my children,' she says. 'The pay is good.' She has little education, and had no job before this one.

So when I asked what she told her priest, and what she would tell her children when they were older, she looked at me in disbelief and swore. 'Well, I ask for forgiveness at confession,' she said, 'but I don't tell him what I do. I'm scared – scared that the families of those I have killed will come after me, scared that my children will eventually discover that their lives were supported because I killed for money.'

She has one more hit to carry out, she told me, and she wants to stop. But her boss has said he will kill any member of the gang who tries to leave. And perhaps the money, for a woman with few other options, is just too good.

It is hard to link this jittery woman, hugging her baby, with the pictures of blood-soaked corpses found every night in the slums and shanty towns of Manila. But I have to make that mental leap, to keep in mind what she does. Before she left I asked her whether, if the price was right, she would shoot me. She giggled: 'You're a foreigner. Why would I kill you?'

But she didn't say no.

Howard Johnson – Philippines – Trolled – 07/02/19

The explosion in social media usage, and the changing news landscape to be found there, throughout the 2010s would raise serious challenges to 'traditional' news outlets – however carefully they'd built their reputations for trustworthiness in print, TV or radio. After the 2016 US election, talk of 'fake news' from the 'lying press' with its 'agendas' and 'narratives' was everywhere – and direct threats to reporters became almost routine. In the Philippines, Howard Johnson saw the line between online trolling and outright incitement to violence blurring . . .

'If I ever see him in the street, I will punch him in the face!' read one of the comments about me.

'The only one in danger will be this reporter, if he doesn't stop the lies and biased reporting he is spreading,' went the next.

I continued scrolling . . .

'Howard, watch your back!' – there was a skull and crossbones emoji next to that one.

'Son of a bitch!'

'Whore-nalism' – that's a fusion of the words 'whore' and 'journalism'.

'You're a *Press-titute!*'

I think you get the idea.

By now the comments were flooding my social media accounts. So much hatred and anger in their fingertips. But I continued reading in morbid fascination. I could have shut down my laptop and walked away. But this was my first experience of a sustained troll attack and I wanted to try to understand it.

I'd just made a documentary about Rodrigo Duterte featuring some of his critics. They accuse the president of attacking anyone who tries to check his powers – including the judiciary, the legislature and the media.

Whenever I posted about it online – on Facebook or Instagram – angry comments followed. Many were littered with spelling mistakes and grammatical errors. Some were written entirely in bold CAPITAL LETTERS to appear more urgent and aggressive.

While threats of violence from faceless trolls are just words, the Philippines can be a dangerous place. Thirteen journalists have been killed here since 2016.

I wondered if the abusive comments were being written by the president's hard-core supporters. After all, this is a man who was democratically elected by 16.5 million people – even after he'd warned them: 'Don't vote for me, because it will be bloody.' Or maybe they were being churned out by paid-for trolls? In 2017 Mr Duterte admitted paying US$200,000 to orchestrate online attacks on his critics during his presidential campaign.

In the documentary, I'd explained how many journalists here say a culture of trolling is being used to stop critical reporting of the Philippine government. To 'bludgeon the media into silence' one reporter told me.

Rappler, the Filipino news website, is a popular target. Its forensic coverage of the president's war on drugs has angered Mr Duterte's supporters. Up to 2,000 comments a day are posted on social media accusing the website of bias. The editor in chief, Maria Ressa, has received online rape and death threats.

Then in July 2017, Mr Duterte used a televised national address to falsely accuse the company of being funded by the CIA. Then came the lawsuits. The company and its CEO are facing charges of tax evasion and illegal foreign ownership, all brought by government departments. Rappler says the charges are politically motivated and aimed at shutting down the website. In 2018 Maria Ressa was named by *Time* magazine as its 'person of the year' for 'holding power to account'. If found guilty, she could face fifteen years in prison.

Back to my own trial by social media.

'Faggot . . .'

'You're not welcome here – get out of my country!'

'Biased, biased, biased.' Funny, that word pops up a lot.

In just two weeks, I received more than 500 comments and messages. A few looked like they'd come from bot accounts, but most appeared to have been written by real people.

The attacks even spread to my posts about completely unrelated stories. About six months ago I'd shared the tale of Tisay, my Filipina rescue dog. When I found her, she was flea-bitten and emaciated, begging for scraps. But after two months of tinned food, medicinal baths and general pampering, she was transformed – full of life and happiness. I'd even posted a collage of before-and-after photos of her.

Underneath them, someone had written: 'Presstitutes like you are not welcome in the Philippines – I'll hunt you down one day!'

These posts seemed organised. The same language and themes repeated over and over. Pre-prepared infographics attacking the opposition or declaring support for the president punctuated the comments. An animated image of Donald Trump angrily jabbing his finger at a crowd was accompanied by the words 'YOU ARE FAKE NEWS' in capitals.

It went on for two weeks, then – nothing. The abuse abruptly stopped. It was as if a switch had been turned off. The trolls had moved on, no doubt busily abusing someone else.

For the record: I didn't reply to any of the comments. This is my response.

Rupert Wingfield-Hayes – South China Sea – Disputed Islands – 20/06/15

At the level of strategic analysis, there's one overwhelming story in East Asia: the unstoppable rise and expansion of the People's Republic of China. The blistering growth of the Chinese economy may have slowed in the wake of the global financial crisis, but there was no scaling back of the country's geopolitical ambitions. For the most part, Chinese foreign policy stresses peaceful economic cooperation as the best way for the country to pursue its interests; but there are some places where a more direct, and far more militarised, approach is preferred. Rupert Wingfield-Hayes saw how it's giving new meaning to 'facts on the ground' – even at sea.

I suspect that until a few weeks ago the South China Sea was not a place most people around the world gave much thought to, or any thought at all. The Spratly Islands even less so. Unless you're a China geek, you've probably hardly heard of them. But google the Spratly Islands today and you will find a sudden deluge of articles, all proclaiming the same sort of sentiment: 'China and the United States are on a collision course and it could end in war!'

Could it? Probably not – not any time soon anyway. But what's going on in the South China Sea is still very significant.

It's all about China, or rather China's intentions.

For the best part of two millennia, China was *the* dominant power in Asia. But then along came European expansion, the Industrial Revolution and the arrival on China's shores of the Portuguese, the Dutch, the French ... and, eventually, the British. China was brought to its knees, carved up, its palaces burned, its people hooked on opium supplied by Britain. Then came a revolution, a civil war, a world war, another revolution and thirty years of Maoist madness.

Now finally China is emerging from those two centuries of chaos. It is once again wealthy, united and strong. None of us knows what that will mean. One reason is that China's secretive Communist Party leadership never tells anybody its intentions. And so we are left to read the 'China tea leaves', look at what China is doing and try to work out its intentions.

That brings me to the South China Sea. The southern part of it, close to the Philippines, is dotted with treacherous coral reefs, rocks and sandbars – only a handful big enough to be called islands. China, Vietnam and the Philippines have been quarrelling over who owns them for decades. But last year there was a sudden and dramatic change.

Aerial photos taken by the Philippine navy showed a fleet of dredgers anchored off one of the Chinese-controlled reefs. They were seen pumping millions of tonnes of material on to the reefs to form an artificial island. A media frenzy ensued, in which I played my own small part. I think I have a fair claim to being the first Western journalist to see these strange new Chinese islands with my own eyes.

Last July I set out on a Filipino fishing boat to try and find them. One morning, ploughing through a heavy swell 300 nautical miles off the Philippine coast, we suddenly saw land ahead where my chart said there shouldn't have been any. Even the latest Philippine navy flights had not detected any work on this particular reef. But there it was: a brand-new, yellowish piece of land, at a place called Gaven Reef.

This year China's work on the islands has accelerated dramatically. More than 2,000 acres of new land has been created on six reefs. In April fresh photos showed the outlines of a runway being laid on one. So what is China up to? Some pro-Beijing scholars have tried to claim the islands are for both military and civilian use. There will be lighthouses and shelters for fishermen, they say.

Well maybe, but Beijing is not spending billions of dollars on huge land reclamation hundreds of miles from its own coast to help fishermen. These islands are military and strategic. China is alone in claiming the whole of the South China Sea. Now it is creating 'facts on the ground'. That runway is not for tourist flights.

A few weeks ago, a US surveillance plane deliberately flew close to the new islands. The crew recorded the immediate and angry Chinese response. 'Foreign military aircraft, this is Chinese navy,' the operator announced. 'You are approaching our military alert zone. Leave immediately!' The warning was repeated with growing irritation until the radio operator was left spluttering, 'You, GO!'

And so the question arises again: What is China's intention? Does it really intend to use these islands to enforce its claim to the whole of the South China Sea? Will it really try to exclude other militaries, including the USA from entering these waters and airspace? If so, there is going to be trouble.

China

Juliana Liu – My Country Cousin – 18/06/11

Much of the news coverage of China in the last decade comes stuffed with intimidating statistics. Staggering sums of money were being invested in infrastructure projects overseas. Imports of raw materials to power China's industries ran at record levels. The world's largest building boom never slowed: factories, shopping centres, and millions and millions of houses mushroomed. The grand sweep of Chinese history now included a great internal migration, of millions moving from the countryside to the cities and from poverty to the global middle class. Juliana Liu in Shanghai told the kind of intimate family story that reflected the story of an entire generation.

I see him on the corner of Shanxi Road, near Shanghai's main domestic airport. It's been more than twenty years, but I recognise my cousin Shen Jun instantly. The same floppy hair. The same moonlike face with eyes that narrow into happy slivers. And the same voice in a familiar Hunan accent calling out Xiao Mao, or Little Hairy Head, the nickname of my childhood.

We Chinese rarely hug, at least not each other, so Shen Jun and I clasp hands in greeting. Then he introduces a woman standing nearby wearing a fashionable black Adidas tracksuit. This is Sao Sao, your sister-in-law, he says.

Hellos exchanged, Shen Jun and Sao Sao take me to a nearby Kentucky Fried Chicken. This is China's favourite fast-food chain. It arrived on the mainland in 1987, the same year I left. Since then, it has become an aspirational middle-class hangout, a popular place for first dates and, for me now, family reunions.

Shen Jun and Sao Sao tell me they have lived in Shanghai on and off for years, working as migrant labourers. He is a security guard at a posh residential compound. She works at a supermarket. They each make less than 200 pounds a month. It's an absolute fortune in the countryside, where they come from. But here in Shanghai, China's most fashionable and Westernised city, it doesn't go very far. That explains why we're meeting at KFC instead of at their home, a tiny space of just a few square metres with no air conditioning or heating. The Chinese value face more than anything, even money. My seeing their modest dwelling would have meant loss of face.

After ordering, we exchange family gossip. One of our cousins seems to be doing well as a salesman. Another is looking for a job. The conversation is awkward. Even though I still speak the dialect of rural Hunan where Shen Jun was born and raised, and where I spent part of my childhood, it's difficult to bridge the enormous gap between us.

My cousin is barely scraping by in a city that places migrant workers at the bottom of its social ladder, while I am living a life of international opportunity few Chinese would have been able to fathom just two decades ago. That's because, in the 1980s, my father seized the window provided by China's reform and opening to learn English and emigrate. Shen Jun's father, my uncle, stayed on at the family farm, planting rice as our ancestors had for generations.

My cousin was not educated beyond his early teenage years. In today's China, a lack of education and connections means a life of hard labour, either in subsistence farming or as hired muscle. And the more we talk, the more I see that many things about my cousin are no longer the same. The carefree prankster who terrorised my toddler years has become a world-weary man juggling life as a migrant worker and absentee father.

He married young, to a local girl. They had a son. Then, as waves of ambitious people had done before him, Shen Jun left for booming Guangdong province, near Hong Kong, to try his luck as a factory worker. The distance contributed to his divorce.

He met his current wife, who is from a different part of China, while working. It's a modern migrant marriage, and they seem happy with each other. Sao Sao jokes that Shen Jun isn't clever or lucky enough to get rich. He smiles in response, saying one's life is determined by heaven, that it's pre-written and impossible to change. But for all his fatalism, Shen Jun is determined to ensure that his son, Wen Long, doesn't follow in his footsteps.

Sounding very much like a guilt-ridden working parent, my cousin says it's impossible to make a living in the city as well as take care of matters at home, which is a day's journey away. That's because even though Shen Jun and Sao Sao work in Shanghai, they are considered rural residents under China's archaic household registration system. It bars them from social services, including health care and education, in Shanghai. So, like almost all the children of migrant workers, Wen Long is being raised in the countryside by his grandparents, who are too old and too soft for discipline.

He gets poor grades in school, and seems destined for a life of manual labour. So I suggest Wen Long should consider training to become a hair stylist, a mechanic or, perhaps, a train driver.

All too soon, it gets late, and Shen Jun must start another night shift guarding the homes of rich locals and foreigners. I insist on giving them a ride by cab. But, proud as they are, Shen Jun and Sao Sao insist on going back by bicycle, the way they came.

Lucy Ash – The Love Hunters – 26/09/13

The tumult of China's emergence from communism was mirrored by drastic changes to family life. As most of the country drifted away from Mao-era sobriety, a number of the social ills the Communist Party had tried to wipe out (bride-trafficking, prostitution, powerful men keeping second or third families) re-emerged. Parents worried that a generation of only children, born as the one-child policy was strictly enforced, would end up as unbearably spoilt 'Little Emperors'. Life grew harsher in many villages – now populated almost exclusively by children and the elderly, as anyone of working age had headed to the cities. And with a gigantic (but undetermined) number of 'missing daughters' – that is, a shortfall of millions of girl children across the nation – the marriage market grew ruthless and more unequal. Lucy Ash found that wherever you were on the social pyramid, romance could be hard to find without finance. At least those at the top could call on professional assistance.

Clutching his iced coffee, Peng Tai strolls down the street and disappears into a shopping mall. Halfway up the escalator, he surveys the scene below.

'What about that girl in the yellow dress?' I venture. 'Uh-uh,' he says, 'too short.' And the girl in micro shorts with long legs coming out of a shoe shop? 'No way!' he says. 'Too skinny. I'm looking for girls with nice skin, nice white skin. They should be not too thin and not too chunky, with a normal way of walking.'

A minute later he sidles up to a twenty-something fashionista trying on perfumes. 'Are you single?' he coos. 'Looking for love?' She quickly shakes her head and walks away. He gets the same brush-off from a few more young women; some look embarrassed, others merely impatient.

Peng Tai rejoins me by the entrance and sucks dejectedly on his drink. 'The girls shouldn't look angry,' he tells me. 'We certainly don't want one with a sour face!'

Suddenly he spots his prey: a very young girl in a sleeveless top with platform heels. Cautiously he approaches her, all smiles. She listens wide-eyed to his opening gambit and looks intrigued as he takes down her details. Then, iPhone in hand, he moves in for the kill and snaps her portrait in the doorway of the mall.

Peng Tai has a quota: three suitable girls a day. He's what's known as a love hunter, working for the Diamond Bachelors Agency. This Shanghai outfit has hundreds of wealthy single men on its books looking for wives. The joining fees range from $18,000 to $1.8 million a year, depending on the level of service required.

Peng Tai and dozens of others like him remind me of fishing trawlers, scooping up whatever they can in their nets. But their catch must then be specially prepared for the customer – and China's billionaires are a notoriously fussy bunch.

One client insisted on a girl who looked identical to Zhang Ziyi, star of the hit film *Crouching Tiger, Hidden Dragon*. Another, a real estate tycoon, paid the agency to search nine cities and interview 10,000 girls to find Miss Right. Of course she had to be stunning. But she also had to be between twenty-two and twenty-four years old *and* have a master's degree from one of two top universities in Beijing or Shanghai.

Peng Tai's work is unforgiving, but the rewards can be huge. Top love hunters can earn bonuses worth tens of thousands of dollars. But I wonder if he ever worries about treating women as commodities. 'I don't care what the girls think,' he says. 'This is my job and we are providing a much-needed service.'

While China's wealthiest men contract out the search for a spouse and have plenty of choice, at the opposite end of the scale, some have no choice at all.

Firstly, the country's inexorable economic rise has put marriage way out of reach for many men. In the early days of the People's Republic, a bride would be satisfied with the so-called 'three rounds' – a bicycle, a watch and a sewing machine. In the 1970s, grooms were expected to provide a fridge, a TV and a washing machine. Now it's a car, a good salary and real estate. One young engineer I met in a Beijing park told me that he'd have to save up for 200 years to afford a one-bedroom apartment – and that would be without eating or drinking.

But China's poorer bachelors face an even more formidable obstacle than a lack of cash: the shortage of young women. Many families value boys over girls, as sons traditionally provide for parents in their old age, carry out funeral rites and continue the family line. When ultrasound technology came to China in the Eighties, it led to sex-selective abortions and a dramatic gender imbalance. The problem was exacerbated by the one-child policy: today 118 boys are born for every 100 girls.

Men in the poorest, least-developed areas suffer the most from the skewed sex ratio because of another inescapable trend in modern China – mass migration. In the past decade 300 million have moved from the countryside to the cities, and for many young women it's a one-way ticket: they marry up and never come back home.

There are 700 people in the village of Tanzhen, in mountainous Guangxi province, and sixty of them are unmarried young men. Sitting in his courtyard, thirty-year-old Wei Tianguang tells me virtually all the eligible young local women are working in factories on the coast. He admits he has never had a girlfriend. 'I try not to think about it,' he says, lighting his umpteenth cigarette. 'And when my mum nags me, it just makes me depressed.'

Does he have any special requirements, I ask?

'Oh no,' he says. 'I would marry any woman prepared to live here with me – anyone at all.'

John Sudworth – Bo Xilai, Petitioners and Justice – 31/08/13

Is China a country of laws? It certainly believes in law and order – enforced with considerable state money and manpower, when it's considered necessary. But amid all the exhilarating news about its economy during the 2010s there was also a steady stream of stories about corruption and influence-peddling, and the apparently hazy distinction between government-funded and private business interests. The trial of the popular former Communist Party head in Chongqing, Bo Xilai, began in 2013, and drew attention to the way the Chinese justice system really works. Bo had once been a party strongman with unusual charisma and approachability, who managed to combine hard-left social policy with running a thriving industrial city. But there was disquiet about his alleged local cult of personality and his all-out pursuit of organised crime bosses. At first he was criticised for overreaching, but later on he was officially accused – and convicted – of corruption, stripped of all his assets and jailed for life. Reporting on this 'trial of the century' spurred John Sudworth to think about how much faith Chinese citizens really have in their courts.

It was the crowd I noticed first and then the object of their attention. An elderly man had climbed onto a tall concrete fence post, and a group of uniformed police officers and plain-clothed thugs were trying to coax or threaten him down.

But the man, grey-haired and, I'd guess, at least sixty years old, was enjoying his moment in the spotlight, having caught the attention of a few international camera crews. He continued his impressive monologue while the men immediately below leapt at his just-out-of-reach ankles. His final defiant act was to pull down his trousers before being manhandled off his pillar and bundled away. The discussion later among the press pack was how best to describe his underwear. 'Billowing' was perhaps the most appropriate offering I heard.

It was an ostentatious attempt to court publicity, but by no means the only one I saw during my time in the city of Jinan. On almost every corner people would sidle up and push documents into my hands. Claims of abuse by the authorities. Photographs of injured relatives or neighbours. Evidence of wages unpaid. Details of homes torn down or land illegally seized by corrupt bureaucrats in collusion with the police.

These approaches are familiar to every foreign reporter based in China. It's just that Jinan was hosting one of the most closely scrutinised show trials China has ever staged, and Bo Xilai's moment in court had brought the petitioners, as they're known, out in force.

Petitioning in China dates back to imperial times, when any subject was entitled to take his or her complaint to the emperor. It rarely got anywhere – and it doesn't under the communists now. As few as 2 per cent of attempts are said to end in any form of success. But regardless of the odds, the number of petitioners has grown over the past few decades.

China's booming economy has produced plenty of winners, but there are losers too. The forced sale of farmland for factory and real estate development has long been a key driver of Chinese economic growth and many petitioners claim to have received no compensation. They don't trust or simply can't afford the Communist Party-controlled court system and so, estimates suggest, some ten million petitions, possibly more, are now lodged every year.

The necessary documents, outlining the particular complaint, are submitted first to local petitioning offices, set up across China. But complaints can be taken all the way to the top ministry-level office in Beijing. And many persevere in spite of the well-documented stories of petitioners being stopped from reaching the capital – beaten or illegally held in so-called black jails – by local governments keen not to have their records blemished by too many complaints.

I recently spoke to one fifty-seven-year-old man who'd been to Beijing to petition more than thirty times. He'd been locked up in a labour camp for the best part of two years for his efforts to seek justice in a medical negligence case. The thought that foreign journalists might be able to help with such complaints is, I've often thought, an indication of the desperation of the petitioners' cause.

There's a view – from the Chinese government, certainly, but sometimes echoed by foreign executives headquartered here in Shanghai – that calling for democracy misses the point. China has achieved so much in the past thirty years, raising living standards and dragging millions out of poverty. People are freer to travel, to some extent to speak and to grow rich than ever before.

If democracy merely means a tick in a box every four years or so, then perhaps their argument holds. But if it means an accompanying system of rights enforceable through the courts, then you could argue that there is a growing hunger for that in China. Even some insiders are urging reform in that direction, fearing that, without it, the simmering social tensions here could eventually reach breaking point.

The trial of Bo Xilai is being sold to the Chinese people as proof that no one is above the law. As it ended and the heavy security presence around the court evaporated, a new batch of petitioners emerged onto the streets. Where had they been, they were asked? These were the truly unlucky ones. They'd been illegally detained in advance, they said, swept off the streets and locked up for the full five days . . . so they wouldn't spoil China's chance to demonstrate the strength and fairness of its legal system.

Carrie Gracie – Xi's Legend and His Cave – 17/10/15

After Xi Jinping became the seventh president of the People's Republic of China in March 2013, official efforts to promote and glorify his leadership intensified. Much was made of his life experience and his ease with the masses. Not for the first time, or the last, Carrie Gracie found the message being driven home with a somewhat heavy hand.

In general, the president's spin doctors do a very slick job of presenting him as a man of the people. He tours leaky back-alley homes, ducking through washing lines and wearing no face mask ... the message that this is a leader prepared to breathe the same polluted air as you. He talks to his public in earthy prose, telling students that life is like a shirt with buttons, 'where you have to get the first few right or the rest will all go wrong'. He queues up in an ordinary dumpling shop for lunch and pays for his own meal. Message: he is neither greedy nor showy.

It's all clever political signalling, of course. Behind his smiling mask, Xi Jinping is a ruthless operator. You don't rise to the top of the Chinese Communist Party without being a consummate political player, and Mr Xi has spent a lifetime playing.

The president was in the car in front of us. And it was stopping at a police barrier. We'd arrived at the heart of his creation myth. Nearly five decades ago, the fifteen-year-old Xi fled from the chaos of the capital to this bleak and breathtaking landscape of yellow canyons, caves and mountains. The contrast with life in Beijing must have been even more extreme in those days.

Especially for Xi. He had actually lived two lives by the age of fifteen. In the first his father was a hero of the communist revolution, so Xi spent his early years as a so-called 'red princeling' enjoying a privileged upbringing. But that was shattered by the civil war that an increasingly paranoid and vengeful Chairman Mao inflicted on the party elite in the 1960s. Xi's father was jailed, his family humiliated. One of his sisters died. Without parents or friends to protect him from the murderous red guards dispensing summary justice on the streets, the teenage Xi lived his second Beijing life, dodging death threats and detention ... until he came here.

Millions of Chinese city kids were doing the same thing. Mao had decreed they should spend time in the countryside, learning from the hard life of the peasants, and Xi Jinping says he did learn.

The spin doctors have turned his teenage trauma into triumph. This village has become a shrine to its most famous son, a vital part of the president's image. 'I left my heart in Liangjiahe. Liangjiahe made me,' he likes to claim.

There aren't many twenty-first-century leaders of whom you can say that they lived in a cave and made it as a farmer before clawing their way to the political summit. But in control-conscious China, those facts could not possibly be allowed to speak for themselves. So I was marched round a museum extolling the good deeds Xi did for his fellow villagers; whenever my attention flagged, a gushing guide stepped forward to fill in narrative gaps, and I soon realised that what I'd mistaken for phalanxes of Communist Party pilgrims were actually propaganda officials. Also keeping an eye on me rather than the museum exhibits: a liberal sprinkling of plain-clothes police.

Why the paranoia? Why does the history have to be sanitised, all trace of real personality expunged? I wasn't looking for revelations of youthful depravity or character flaws. But everywhere I turned, the memories were so carefully crafted it was hard to work out whether any of them were real. And all the while, the propaganda chief's glassy pallor worsened. Eventually he asked me to sign a document promising that every word the BBC said about President Xi would be positive. He blanched when I said I couldn't. It might be his job to promote the personality cult, but it wasn't mine.

The great irony is that under Chairman Mao, Xi Jinping and his family themselves suffered from strongman politics. After the tragedy of the Cultural Revolution, the Communist Party resolved it would never make the same mistake again. Grey, faceless committees ran China for the next forty years. But now the strongman is back.

For us it was time to leave the caves. The propaganda chief had started threatening to confiscate our recordings. President Xi may have left his heart in Liangjiahe, but I didn't want to leave all the material for my documentary there. That night we made a sudden bolt, driving several hundred miles to an airport from which we could get our work out of China. Strange exploits, when you consider that the cave years seem the most positive chapter of Xi Jinping's life, even without persuasion and threats from the propaganda department.

Celia Hatton – Corruption Trials Have Them Running Scared – 10/03/16

Bo Xilai's trial showed which way the wind was blowing: even party high-ups could be brought low in an instant in Xi Jinping's China. Once, this sort of disgrace might have been inflicted on ideological grounds – or it might even follow a mere rumour, or a tactless remark. But in the 2010s, the denunciations usually followed accusations of financial wrongdoing: fraud, theft, corruption. Even members of the Politburo were investigated and charged – and juicy details about their mistresses, fat portfolios of shares, fleets of fast cars or spoiled children came spilling out during their trials. After a slew of big anti-graft prosecutions, Celia Hatton could feel the chilling effect in the capital.

I had already handed her my credit card, but Jane wanted to chat. She was taking her time ringing up the sale. When we first met, a decade ago, she had just landed her job in one of Beijing's touristy pearl stores, known for selling earrings and necklaces that cost a fortune elsewhere.

For years, whenever I visited, the store was crowded and chaotic. Mountains of pearls would be piled high on tables; there were so many that loose pearls rolled around on the floor. No one had time to sweep them up – they were too busy making money.

But now, the store was empty. And not just this one. All the pearl stores in this area had fallen silent.

Unsurprisingly, Jane was bored. She was also eight months pregnant. Instead of her usual uniform suit, she wore the stereotypical outfit for expecting Chinese women: a pair of corduroy overalls with a teddy bear face on the front.

'Where are all your customers?' I asked. In the hour I'd been in the shop, no one else had entered.

'Look outside,' she sighed, gesturing to a distant window. 'The tourists have disappeared. They're scared of the pollution ...

'But that's not the worst thing,' she continued. 'Our Chinese customers used to spend the most, by far. Government people bought our best pearls as gifts. But now, they're all scared of Mr Xi!' she smiled, referencing China's leader, Xi Jinping. 'During the anti-corruption campaign, no one wants to be seen wearing pricey jewellery.'

It's been announced that around 300,000 officials were punished for corruption in the last year alone, with 80,000 receiving 'severe punishments'. The campaign's work takes place internally, so little is known about who those officials are, what they did wrong, or what happened to them as a result. But the anti-corruption drive has other, more obvious consequences: spending on flashy luxury items like watches has dropped. Jane's pearl store fell victim to this. Gambling has taken a hit too. The Chinese enclave of Macau, the Las Vegas of the East, has seen profits drop 20 per cent in the past year – all due to the absence of worried officials, it's said.

But that doesn't mean the Communist Party has cleaned up its act. Far from it. 'I don't think this is making a big dent in how many officials are breaking the law,' one China-watcher tells me. 'That's not what this is about.'

People outside of China are misunderstanding this campaign, she explains. Ultimately, Xi Jinping isn't really trying to groom law-abiding officials. No, he's trying to create legions of obedient officials to implement his policies. This is a bid to ensure loyalty at every level of the Communist Party, from the upper echelons in Beijing down to the most remote villages and stretching right across all the government's financial empire, too. China's state oil companies, the military and the big banks have all been subjected to corruption inspections – even the corruption bureau itself has its own internal inspectors.

Every 'rogue official' who might have been trying to carve out a personal empire by setting up a network of bribes and kickbacks has been taken out and replaced with someone promoted from within: someone loyal to President Xi. A certain amount of corruption and graft is tolerated, as long as the ultimate priority, party unity, stays intact.

In this environment, all sorts of things are still going on behind closed doors. High-end lingerie sales are soaring, the logic being that communist cadres will still spend money on private luxuries. Well-connected friends tell me that 'secret' restaurants are viewed as a necessity now. On the ground floor, they look like shabby teahouses, but banquets are served on the upper floors. Yes, these luxuries are still popular. It's just the consumption that's gone underground – quite literally, I found.

In Beijing, I was always puzzled by a very expensive cluster of stores near my home. It was an outdoor shopping plaza, full of the ritziest European labels – and it always seemed to be empty, with almost no one going in or out of the shops. That is, until one day, I entered the underground car park below the plaza. Every space was filled with gleaming cars: BMWs, Lamborghinis, a Rolls-Royce or two. And behind the cars stood the individual elevators. Each store has its own plushly carpeted lift. No need to ever venture outside, into the public eye.

'When do you think the current anti-corruption campaign will end?' I ask a respected expert I know. He laughs. 'It won't end until Xi Jinping feels completely in control,' he says. 'He's totally reliant on his loyal cadres, but everyone else hates him and is just waiting for him to stumble. As soon as that happens, the corruption will come flooding back.'

No need for underground car parks when that happens.

John Sudworth – Xinjiang: Beyond Orwell – 01/12/18

Alongside the national narrative of dizzying economic progress ran another: of ever greater integration and ever more forcefully policed national unity. China could not go off-message – and that applied with particular force in Xinjiang in Western China, home to the country's Uighur minority. This mainly Muslim people traces its heritage to Turkic-speaking groups to the west and has long been treated with suspicion by the Chinese state. The Uighurs have responded with waves of unrest, and at times with violence. After hundreds of

deaths during rioting between Uighurs and Han Chinese in the re-
gional capital, Urumqi, in 2009, Beijing brought in the troops, and
then used other techniques to pursue what it called a 'people's war
on terror'. More recently, camps have been set up which China says
are meant to 're-educate' extremists and help them return to normal
life. Human rights activists say up to a million Uighurs have been
sent there to be brainwashed. John Sudworth went to Xinjiang to
investigate.

'Nothing was your own except the few cubic centimetres inside
your skull,' George Orwell wrote of the experience of living under
the all-seeing eyes of Big Brother.

In the years since the publication of *1984* many readers have
pondered how much of the book's dystopian future has become
a reality. But the totalitarianism it describes is so total – such a
perfect blending of old authoritarian repression and futuristic technological control – that nothing has ever really come close. Until
now, perhaps.

Xinjiang – China's vast, far western region – is the target of the
world's most comprehensive and sophisticated police state. Among
the first things any visiting reporter notices are the ranks of cameras, watching people's every move, every few metres. Many are fitted
with facial recognition technology. DNA samples have been taken
en masse. Mobile phones have to be unlocked and handed in at
police checkpoints. At one of them, I watched as officers plugged
their scanners into the collected phones, monitoring them for illegal
content. The checking is constant anyway. Every Xinjiang resident
has been forced to download an official spyware app. Failure to do
so is an offence.

All of this, China says, helps combat the threat of Islamist terror.
The primary focus of this extreme surveillance are the Uighurs –
the region's main, largely Muslim, ethnic group, some ten million
strong. And Big Brother is not just watching them: he is living with
them.

A programme called 'becoming kin' has sent over a million
Chinese government officials to stay, for regular periods, in

Uighur homes. The officials monitor their adopted families for signs of devout religious behaviour or other indicators of potential disloyalty to Chinese rule. A negative assessment can mean being sent to a re-education camp. Hundreds of thousands of Uighurs, and other Muslim minorities, have already been detained.

Outside one of the biggest known camps, we were stopped from filming. But we were keen to find out what local people knew about it. As our police minders made it impossible to speak to anyone openly, we later called phone numbers in the town at random. 'There are tens of thousands of people there now,' one shopkeeper told us. 'They have some problems with their thoughts.'

It was Orwell's *1984*, of course, that popularised the term 'thought crime'. As the main protagonist Winston Smith sits down to write his subversive diary, he knows he is already doomed.

In Xinjiang, though, there are no dissenting voices left, not even mild ones. Those who simply once studied or defended Uighur culture – university lecturers and prominent intellectuals – have been taken to the camps. China insists that they are places of 'training', providing job skills alongside classroom-based, deradicalisation lessons. But many have watchtowers, barred windows and barbed wire. 'Doublethink', Orwell would call such a contradiction between language and reality.

It was not meant to be this way. By the end of the twentieth century totalitarian regimes appeared to be in full and final retreat. Orwell's novel, even as political satire, was perhaps starting to look a little dated. The Soviet Union had collapsed. Communist China was rapidly transforming itself into a capitalist superpower.

But alongside the rise of the Chinese economy the accompanying political reform, predicted by so many for so long, never came. What arrived instead were the internet and the mobile phone – technologies with a potential for mass surveillance that George Orwell could never even have dreamt of. And the Chinese government has seized that potential like no other.

As *1984* – the novel – winds towards its long, dark conclusion, even those few cubic centimetres inside the skull are shown eventually to belong to Big Brother. So it is too in western China, where thousands of Muslims are undergoing thought control today. 'Orwellian' may have become an overused political term. But in Xinjiang, it has never been more appropriate.

Oceania

Jon Donnison – Australia – Sydney Fires – 26/10/13

Australia's environment can be daunting. Everything about it can seem huge, hostile, extreme: the heat, the wildlife, the distances. With so much of its population crowded near its south-east coast, keeping everyone supplied with enough water and safe from bushfires is an immense challenge for planners. Naturally occurring fires have shaped the landscape for millions of years, and many Australian species have specifically evolved to benefit from them. And ever since the first human settlement of the continent by Aboriginal groups, tens of thousands of years ago, fire has been periodically used to burn off huge tracts and change the natural balance of plants and animals. In a fast-changing climate, the risks are greater: average summer temperatures are climbing ever higher, and the resulting fires more frequent and more deadly. But covering the blazes in New South Wales in 2013, Jon Donnison found plenty of scepticism.

Australia's Blue Mountains get their name from the eucalyptus trees carpeting their slopes. On a warm day the sun heats up the oils in the trees' vibrant green leaves. As those oils evaporate into the atmosphere they give the range a shimmering blue hue. For much of this week, though, the Blue Mountains were more a shade of grey, cloaked in smoke, the air acrid and woody. And there's no smoke without fire.

These have been the most devastating bushfires New South Wales has seen in decades. Tens of thousands of hectares have been burned. Hundreds of homes destroyed. And they're still burning. In the small community of Winmalee about an hour's drive from Sydney, I looked on as Chris Muller stood, head in hands, in front of what was left of her house: pretty much nothing. A once beautiful property, looking out over the bush, reduced to little more than rubble, twisted metal and ash. Many of the houses on the suburban street had suffered the same fate.

Chris's daughter, picking through the debris, tried to cheer her mother up. 'Look, Mum!' she shouted, holding up a few metal spoons and an old blue coffee pot. 'I guess we're all still here, that's the main thing,' Chris said before heading off to talk with her family about how they were going to remake their lives.

A local sports club was transformed into a disaster welfare centre, offering food and advice to those who'd lost their homes. Husbands and wives, sons and daughters could be seen poring over long to-do lists. The first question surely: where to begin?

Of course Australians are used to the threat from bush fires. And there have been much more deadly ones than these. More than 170 people were killed in 2009 in the 'Black Saturday' fires in the state of Victoria. But this year the burning has come unusually early and after unseasonably hot weather. It's still only spring. The full heat of summer, with its even higher fire risk, is months away. And according to Australia's Bureau of Meteorology this follows the country's hottest year on record.

That fact has given these fires a political edge. Environmentalists often like to refer to Australia as the petri dish of climate change, where you can actually see global warming happening and judge its effects. Australia's new conservative prime minister, Tony Abbott, has sometimes seemed reluctant to accept that view. Mr Abbott, who's a volunteer firefighter himself and even gave up his day job to man the hoses these last few days, once described the science behind human-induced climate change as 'absolute crap'.

He's since refined that opinion and now accepts global warming is happening. But this week he was quick to dismiss comments from a senior United Nations official who had linked the fires to the changing climate. 'She's talking through her hat,' Mr Abbott said, adding that fire had always been part of the Australian experience, before listing a host of major fires across the country stretching back more than a century.

The former US vice president and climate change campaigner Al Gore said denying the link between these latest fires and global warming was like claiming smoking didn't cause lung cancer. People like the Nobel laureate are already dismayed that, since coming to power last month, Mr Abbott has pledged to dismantle Australia's carbon tax.

The tax, introduced last year by the previous Labor government, makes the country's biggest polluters pay for the greenhouse gases they produce. Instead, Prime Minister Abbott favours what he calls 'direct action' – giving businesses financial incentives to adopt green technology, if they choose to. Within days of taking office, he shut down the country's Climate Commission, which had been set up to compile scientific research into the issue. All this has put the prime minister at odds with environmentalists. And if the fire season continues as it has started, the heat of the argument over the causes is also likely to intensify.

It's a divisive issue in Australia – but up in the Blue Mountains, most people I spoke to this week tended to side with Mr Abbott. This year's fires were 'all part of the cycle', Col McDonnell told me as he prepared to defend his farmhouse while thick white smoke billowed from the hillside behind him. He said that fires like this had been going on through the ages and that it was all part of life in the bush.

Madeleine Morris – Australia – Extortionate Limes – 23/03/13

Although life in Australia might superficially seem like that in other high-income developed countries, there are huge differences – and not just in the climate. During the 2010s, its vast mineral reserves and proximity to China insulated it from the worst of the American and European financial crisis. The bank collapses, austerity measures and fiscal brinkmanship which filled the headlines in the global North simply didn't resonate Down Under. When Madeline Morris went back for a family visit, one of the simplest economic indicators of all told the story: the price of a regular grocery shop . . .

It was the limes that finally tipped me over the edge. In the sleepy Australian seaside village where my parents live, not that far away from several citrus orchards, I was in the supermarket staring at a sign: '*LIMES – AU$2.25*'. That's £1.50. Not for a bag. Not for a pair. Each. One single solitary lime cost £1.50. Infuriated, I stormed out of the shop, limeless.

'The country's lost it,' I fumed to my mum and dad over dinner that night. 'How can anyone afford to eat?'

'Darling,' my father replied. 'Look around. People here are rolling in money. We live in an unbelievably wealthy country.'

And he is right. In the twelve years since I last called Australia home, it has changed. It was always the lucky country, blessed with fertile land, abundant sunshine and plentiful natural resources. Now, we are more than lucky. We are rich. Bloody rich. So rich that no one blinks an eye at paying as much for a single lime as some of our neighbours in Asia might earn in a day.

You can feel it, just by looking at the small stuff. For example, there is no litter on the streets. Nowhere. I have yet to see a central reservation where the grass isn't well-tended or the attractive shrubs perfectly pruned. It's the cars. I swear there are none on the road that are any older than eight years. They're clean and dent-free and meet strict safety standards.

It's there in the obsession with gourmet food shows: the shiny European appliances in the shiny designer kitchens that now seem to feature in even the most average family home. It's the seriousness about single-origin coffee, made by baristas who get paid £17 an hour (before tips) to bestow their caffeine-laced munificence on their devoted followers.

I don't mean to sound flippant. Of course there is poverty too, and the gap between rich and poor is growing. But the overall sense I get is that this is a country that can afford to be worried about the small stuff, because the bigger things – food, shelter, employment – are pretty much taken care of.

Australia was one of the few nations to come out of the global financial crisis without a recession. It was down to the prudent economic management of the government at the time, but also largely because of the huge mining boom this country has been riding for nearly a decade. The world, especially China, wants what Australia has in the ground – and has been willing to pay for it.

It feels to me, a long-lost daughter, that the country has been irrevocably changed. My parents' village used to be inhabited by retirees and fishing families. Now, we share the one pub with hundreds of mine workers, who come for their days off to burn money on bottles of spirits and the newly installed slot machines. Their driveways are stacked with fishing boats, jet skis and monster trucks – all the big boys' toys.

We call them 'cashed-up bogans', which roughly translates as 'rich chavs'. Plenty of money, not much sense. It's a term my middle-class tribe uses disparagingly to make us feel better about being educated, but comparatively poor. I am not the first privately educated Australian university graduate who now wishes she'd done a truck-driving course instead. Sure, I might have been bored at work, but at least I would have been able to buy a house.

The other day I asked my taxi driver, originally from Turkey, if he thought Australians were rich. He looked at me as though I was stupid. 'We are living in the lap of luxury here,' he said, gesturing to the blue sky and the magnificent city skyline. So I asked him if he thought Australians were happy. This time, he sighed. 'When I was at school my teacher asked me who had to work harder, the poor Africans, or the rich Americans,' he began. 'A lot of us said the Africans, but my teacher told me no, it was the Americans. They were always working to find ways to pay for their lovely life. Australians are the Americans now.'

It makes a lot of sense. As a country, we are richer than we could have ever imagined twenty, or even ten, years ago. But we're more anxious too: worried about our non-existent public debt, worried about what we'll do when the mining boom is over, worried about how we're going to pay for our next overseas holiday, because that's what we've all come to expect as normal.

And me? I'm especially worried about how to make the limes on my newly planted lime tree grow, because I sure as hell won't be buying them in a supermarket any time soon.

John Pickford – Tonga – China Reaches Across the Pacific – 27/07/13

Not much news from the small island nations of Oceania makes it to the headlines of global newspapers or the top of international TV news bulletins. But these are not ahistorical, untouched 'paradise islands': they have plenty to worry about. The age of the smartphone means that images of the storms, eclipses or tsunamis they weather can now be shared with people across the world; illegal drugs have been found on many of the islands; and there's gnawing, long-term concern about the fate of many of the lower-lying island nations if sea levels rise. Some of the smaller countries have petitioned the UN in advance for their people's rights to settle elsewhere en masse in future; they might soon be the advance guard of a global wave of climate-change refugees. But for the moment, John Pickford, who regularly travels between these dots in the vast Pacific Ocean, observed one visible trend above all others.

The domestic airport in Tonga's capital, Nuku'alofa, was eerily quiet. The friendly café, where three years ago I'd enjoyed the delicious locally grown coffee, had gone. And no sign of my plane.

Waiting with me in the shade for something to happen was a young Chinese man, booked on the same flight, returning to his wife and grocery store on the island of Hapa'ai. We started chatting and during our long wait I learned quite a bit about Johnny Wang. How he came to Tonga from Shanghai, where his two young children have stayed with his mother-in-law. He sends remittances and visits them during Chinese New Year. How he enjoys his new life on Hapa'ai, 'very quiet, very peaceful', he says, 'and easier to make money than in China'.

We started talking Tongan airline politics. How the Chinese government's gift of a new passenger plane for inter-island flights had prompted the New Zealand-based company that had been providing a domestic air service to pull out, saying it couldn't compete with a subsidised airline.

The Chinese-manufactured MA60 aircraft has been delivered. But no Western government has given it a safety certificate, and New Zealand has just announced it's suspending a multimillion-dollar aid package to improve tourist infrastructure in Tonga until these safety concerns are resolved.

As the whale-watching season gets under way on outlying islands such as Hapa'ai and Vava'au, the flights fiasco is playing havoc with the tourist economy. It's also stirred up concerns among some Tongans that Chinese aid – generous, visible and impressive as it's been in recent years – can be a mixed blessing. Urban roads laid without proper drainage so nearby houses get flooded when it rains. Grandiose public buildings that are poorly adapted to a tropical, oceanic climate, making them costly to keep cool and maintain.

Cabinet minister Clive Edwards chose his words carefully. 'Some of the buildings they've put up', he told me, 'are a disappointment.' I was sitting with him in his office in the Ministry of Justice. 'Including this one?' I asked. He laughed, a deep, Tongan, guttural laugh.

On islands across the Pacific you'll find very visible evidence of Chinese 'big project' aid: sports stadiums and parliament buildings, government offices, police stations. It's their fish, potential mineral resources and votes in the UN that make these small Pacific nations of growing interest to Beijing.

In Tonga, China's visibility is all the greater because of the several-thousand-strong Chinese minority. There is a degree of tension between the two communities. From Tongans you hear phrases like 'the Chinese are bad drivers' and 'the Chinese are everywhere'. What is not in doubt though is the key role the Chinese play in the business sector, especially in the little family-run grocery stores, the Fale-Kaloas, as they're called.

The Tongan extended family runs on the principle 'to each according to his need, from each according to his capacity'. But if you're a Tongan trying to run a Fale-Kaloa, as local shipping agent Fine Tohe explained to me, and a relative turns up to tell you that his grandfather's died, there's going to be a feast and you're going to have to hand over all the chicken in your shop, you're not going to last long.

The plane did finally get me to Hapa'ai and a day later I was cycling across a causeway between two small islands. There's repair work going on and a Chinese foreman in charge. I heard an engine revving behind me and a delivery van rumbled past. I noted how clean and new it was compared with most vehicles on the island. It stopped on the other side of the causeway and a head leaned out of the front window to grin at me. It was Johnny Wang, my companion from the plane. I was delighted to see him; his pale, smiling face suddenly looked so exotic on that lush green atoll beside the clear blue sea. Then I remembered I needed some toothpaste and new batteries for my torch, and I realised that I felt reassured by Mr Wang's presence. I knew he'd have what I wanted.

Twenty minutes later I was inside his shop. It's an emporium compared to the local competition: tools, footballs, rice, tinned food, nappies. And if he's asked for something he hasn't got, he says he'll try to get it, rather than just smiling and shrugging.

As I cycle back with my toothpaste and batteries, I realise that whatever the future role of China in Tonga, the niche Mr Wang is filling on Hapa'ai will be his to occupy for a long time.

Part VI:
THE AMERICAS

USA

David Willis – The Horror of Health Insurance – 07/11/09

One of the deepest political faultlines in America is the gulf between those who worry about the expense of getting sick, and those who don't have to. President Obama insisted that it was unacceptable for tens of millions of American citizens to be without medical insurance, 'living every day just one accident or illness away from bankruptcy'. In Los Angeles, the BBC's David Willis certainly felt uneasy when he discovered that he too was among the uninsured masses.

Isn't it amazing how, within a matter of minutes, fate can build you up almost beyond recognition, only to deliver a well-aimed slap across the backside? Going through the mail last week an envelope marked 'US Immigration Service' contained – ta dah! – a laminated piece of plastic confirming that my application for permanent residency had finally been approved. Holding my green card, I was just about to break into an interpretative dance of celebration when I spotted another envelope, and the euphoria evaporated faster than a fart in a fan factory.

The letter was from a company which oversees the BBC's health insurance plan. I was confronted with the news that my coverage – extended after I left the corporation to go freelance – had come to an end. This was hardly unexpected, but seeing it in black and white filled me with horror. Now I was on my own: it was me against the system and I had a feeling things were about to get ugly. Because when it comes to health insurance, the Americans could teach the British a thing or two about bureaucracy.

It's difficult to overstate how vital health insurance is in America. Find yourself in the emergency ward, strapped to machines which go 'bip' and surrounded by doctors who've been called in just to deal with you, and if you don't have coverage, the chances are you'll be paying for your visit until the end of time. Even if you're a picture of health, all it takes is a freak accident and you're toast. An expatriate friend of mine spent a week in hospital after flying a small plane into the sea. His bill: $95,000. He told the cashier he'd come in for treatment – not to buy the hospital.

Being uninsured was especially worrying for me because I am ... well ... the world's biggest hypochondriac. If I get a headache, I instantly assume I'm haemorrhaging. The longer it continues, the wilder my doom-laden diagnoses become: what if tapeworm larvae are burrowing into my brain? Such fears have led me to have virtually every test under the sun; I've donated blood by the bucketful and enough urine to float a battleship, because I know my body is trying to fool me. Yes, I feel as fit as a fiddle but surely it's just a facade: my system is lulling me into a false sense of security whilst some deadly virus is getting busy in my innards, readying for my untimely demise.

Given the fact that visits to the doctor are therefore a weekly occurrence – I prefer to think of it more as a pastime than an addiction – you can see how distressing the prospect of being without insurance could be.

And so with heavy heart I took on the many-headed hydra that is the American health care system. For some reason there was no way of simply continuing the policy the BBC had in place and paying the premiums myself. So I had to apply anew, as if I had never had coverage before.

I found myself talking to Steve, a chirpy salesman at one of the larger insurance companies, who ran me through the details of one of their policies. Yes, I'd still have to pay to see either a doctor or a specialist, but he'd throw in a prostate exam and a colonoscopy every ten years. By the time he'd finished I felt like a winner on *The Price Is Right*. The cost: $5,500 a year!

Steve sent me a form which delved into my every malady since emerging from the womb. And I was reminded that if there is one thing that health insurance companies absolutely hate, it's sick people. Sick people have the audacity to require treatment, which not only eats into profits, but upsets the accountants' balance sheets. Too much of that and you could completely spoil their day.

Having explained away virtually every cough and sneeze over the course of the last forty-nine years I got to Question 41: Has the person applying for coverage consumed any alcoholic beverage in the last six months? I read that several times, even at one point substituting a different pair of glasses, and no, I wasn't mistaken – it really did say six *months*. Not six days, or six minutes, but six months.

By the time I'd finished the form I had a headache and eye strain, and so I went back and added those to my pre-existing conditions and then sent the form off to Steve. He told me my application would be assessed by an underwriter, which conjured up images of Lloyds of London weighing the fate of the *QE2* – or in my case perhaps the *Titanic*. I wait on tenterhooks to learn whether my application has been approved. The tension is killing me. At my age that's just not good for the blood pressure . . .

Mark Mardell – Tea Party Passion – 09/10/10

Well before the Trump presidency, there was a sense of grassroots in-surgency within the Republican Party – a revolt of blue-collar voters with hard-line ideas about the limits to be set on taxes and the power of central government. They agitated to drive the party rightwards, away from the 'Washington swamp' and its inevitable compromises, and to root out the RINOs – Republicans In Name Only, considered mushy on abortion or fiscal policy – from its ranks. Mark Mardell tried to take the measure of these internal divisions.

A forlorn yellow balloon is trapped by the sky, unable to rise any further. It's stuck, against the deep blue heavens and the fluffy white clouds. Gondolas circle St Mark's Square below. It's a very American illusion. For I am in one of Las Vegas's palaces of dreams, the huge Venetian hotel, where a narcotically vivid sky has been craftily painted on the vast ceiling arching way above my head.

The hotel casino seems busy enough, music blaring and lights glaring, but like the balloon the city's economy nowadays has trouble rising to its accustomed heights. Those of a censorious disposition might find it fitting that this city, devoted to the illusion of sudden wealth, is nestled in the desert embrace of Nevada. This state has the unpleasant distinction of having the worst economy in the USA.

Which is why the leader of the Democrats in the Senate, Harry Reid – a Nevada senator for twenty-four years – may well lose his seat when America votes on 2 November. It's a fate that may befall many of his less illustrious colleagues too. For Nevada is not alone. All over the United States you pass the depressing sight of boarded-up homes and businesses. A recent government report showed the recession ended in the summer of 2009. In every state I've visited I have heard the same baleful joke: 'Shame no one told the economy!'

'It's the economy, stupid' is perhaps the most overworked cliché in politics. But it still counts as an insight here because politicians, particularly those on the right, have often focused on cultural issues – abortion, guns, homosexuality, evolution, Islam – instead of the bread-and-butter ones.

But the Republicans may capitalise on the mood of economic unease because of the boiling enthusiasm of the Tea Party movement. Its members have a sense of anger, a disgust that they've been let down and a near-absolute distrust of government. And they know it's the economy, stupid.

This vibrant, anarchic and furious ginger group of conservatives has pushed the Republican Party to adopt their people and their policies. One activist did not demur when I compared their tactics to Trotskyites who attempted to take over the British Labour party in the 1980s. 'Exactly,' he said. And when you look at them gathered for one of their big rallies, with their hand-made placards, they can look, well, a bit nutty. (Their opponents, of course, are eager to make them look complete fruitcakes.) But they are more serious than that.

The core of what the Tea Party is about is smaller government, less spending and thus lower tax. But it is big government that is the real enemy. It isn't just the bad policies but a feeling that government here in itself is almost evil, an assault on freedom. It's a great paradox: a movement that boasts of its theoretical love of America and democracy but which hates its real-life institutions. It's not the fairly mainstream economic theories I strain to understand, but the passion, a passion which means that political discourse has become increasingly uncivil, filled with vitriol and abuse.

So why is the Tea Party boiling so hot?

Some say it's racism. Those I've met are not racist, but I do wonder if for some there's a sense of lost superiority. For all their lives there's been a white man in the White House. It's not just that Obama isn't in this image; cool, cosmopolitan, calm and aloof, he does not fit any stereotype of a black person that they know. There is a sense of disconnect from what, in their view, ought to be the natural order.

One woman who told me that Obama was a socialist and her country was sliding into Marxism said that when he was elected president she drew the curtains for three weeks and couldn't answer the telephone. Only the Tea Party saved her. America is changing fast and many Tea Party people don't like the loss of the assumption that white European 1950s America is the norm, the benchmark.

Americans bang on a lot about being American. There is an intense patriotism, a belief the United States has a unique place in the world and that its actions have (just about) always been for the good. There is perhaps today a puzzling lack of self-confidence in this, the richest and most powerful country the world has ever known. Perhaps it is because, for many, it is a country that was born with a purpose, a mission, a dream. And if their particular notion of that dream is questioned, it feels as unnatural and artificial as that Venice sky in Vegas, stopping the yellow balloon from rising.

The British can look far back into their past history, to Alfred the Great, or even King Arthur, but American history has very shallow roots. To the rest of us, America looks robust, but many American conservatives believe they are nurturing a fragile dream. The American novelist of the moment, Jonathan Franzen, perhaps sums it up aptly in his new bestseller when he writes: 'The personality susceptible to the dream of limitless freedom is a personality also prone, should the dream ever sour, to misanthropy and rage.'

Laura Trevelyan – Gun Drills – 08/02/14

'School shootings': two words which should never go together, but have become a cruelly repetitive refrain in news from the USA. After the December 2012 Newtown school shooting in Connecticut, in which twenty students and six teachers were killed, nearly every school district in the nation tightened security. Safety drills were becoming as common as fire drills, Laura Trevelyan, a parent in New York City, reported.

My seven-year-old is a chatterbox, and as the youngest of three boys, he's always keen to be heard. Little in his life goes unreported. Every day has a banner headline. So he couldn't wait to tell me about the safety drill he and his classmates had practised that morning. 'It's in case someone has a bad day,' Ben announced with gravity. 'We huddle with our teachers and keep quiet until the danger outside has gone away,' he said matter-of-factly. 'There's no danger inside,' he explained for emphasis, gazing intently at me.

I smiled encouragingly, while wincing inwardly, and rapidly changed the subject. The thought of my children learning how to behave in case there's a gunman on the rampage is deeply unsettling. But for schools across America, the lockdown drill has become a grim necessity.

Although statistically the number of school shootings in the USA has remained steady since the mid 1990s, they are a disturbing reality of American life. Since September, there have been eleven school shootings – and that doesn't even include shootings on college campuses. In Nevada and New Mexico, twelve-year-olds have threatened their classmates with guns. In Nevada, a maths teacher who confronted a middle-school shooter last November was shot dead.

The most chilling school shooting of all was just before Christmas in 2012. Twenty first-grade students at the Sandy Hook elementary school in Newtown, Connecticut, aged just six and seven, were killed by Adam Lanza, who shot his mother and six adults at the school before taking his own life. The scale of this shooting was so horrifying and the age of the children so heartbreakingly young that it led to a push for national gun control – which ultimately failed to get through the US Congress.

I reported on the aftermath of the Newtown shootings, one of the more painful assignments of my career. Christmas trees for the twenty dead children lined the street by the school – ornaments decorated by friends and family described six- and seven-year-olds who loved football, horses and Santa Claus. The temporary morgue parked near the scene of the shootings was a particularly upsetting sight; the thought of the dead children inside, the same age as my youngest, was hard to contemplate. It was impossible not to wonder how I would cope if it was my child inside that morgue.

Since Newtown, many schools have adopted safety plans. Metal detectors, surveillance cameras and fences have become a fact of life. All three of my children now practise safety drills – and the school is careful to let parents know in advance in case the topic comes up at home. One parent, echoing my own feelings, told me that it breaks her heart to have her boisterous, talkative five-year-old learning to hide from what could be a gunman. Another mother questioned the need for this kind of drill, regarding it as an overreaction to an event that is still very rare.

We all wonder what impact this has on our children and if it makes them fearful. But most parents see the drills as a necessary feature which could save lives. Meanwhile, efforts to bring about national gun-control laws have stalled. At the State of the Union address in January, President Obama mentioned gun control in passing. Last year it was the emotional centrepiece of his address.

I try to downplay the topic with my children. Two of the boys haven't even mentioned the drills they've done at school. It's little Ben who has reflected most on the subject. I tell myself that even though there are almost as many guns in the USA as there are people, in New York gun control is strict and the police are close by. It would be difficult for a gunman to get into a New York school unseen, I rationalise.

'Do you think anyone will ever have a bad day near me?' asked Ben recently, looking up from his elaborate drawing of dragons. 'No,' I said firmly, crossing my fingers and speaking in my most confident and reassuring manner. 'You don't need to worry about that. It will never happen.'

Apparently satisfied with that answer, he returned to his drawing, while I turned away, my eyes brimming with tears.

Rajini Vaidyanathan – Immigration Limbo – 12/03/15

Long before any rhetoric about 'building a wall', fortifying the southern border was a routine demand of nativist US politicians. And while the USA has often drawn on migrant labour from elsewhere, it's also recurrently 'cracked down' on those without papers or permits. The largest mass deportation was probably the infamous Eisenhower-era 'Operation Wetback' of 1955, in which between 300,000 and 1,300,000 Mexican nationals, many of whom also had US nationality, were forcibly removed. Since then, the rigour of immigration policy, and its enforcement, have ebbed and flowed, but during Obama's second term, Rajini Vaidyanathan found there were always individual and family stories behind the numbers.

Court Number 1 is a soulless, windowless room, where dreams are deferred and hopes are put on hold.

Fluorescent tube lights beam down on the suite, on the second floor of a vast office building. A judge sits in front of a circular wooden crest with an American eagle carved into it – a symbol of the country's freedom. To his right is a translator.

In the rows of seats in front of him, a group of immigrants waits. Most are Hispanic; almost all have entered the USA illegally. Desperate to stay, they're here to convince the authorities to let them. These teenagers and young children are just a handful of the tens of thousands of unaccompanied minors who cross the border into America each year.

'Number 819,' the judge calls out. A teenager named Luis stirs from his seat. Stocky in build, and wearing a sports jacket, he's clutching a bundle of papers, one with his photograph attached.

'How are you?' asks the judge as Luis walks to the table at the front of the courtroom.

'Fine,' replies Luis, as he fidgets nervously.

'Will you find a lawyer?' the judge continues.

Luis nods, as the hearing swiftly concludes. He's told to return in August. Delays like this mean months, maybe years in limbo for those waiting to find out if this country can ever become their home.

Luis's story is all too common. He came from Guatemala, walking for ten days to join his father Cesar in America. Like so many children who make the perilous journey alone, he was caught at the Texan border and sent to an immigration centre. From there he was taken to Virginia, to be reunited with Cesar.

Cesar, who is dressed in a striped shirt and battered denim jeans, is in his fifties. As we talk he clutches a faded baseball cap. Cesar belonged to a generation who left their country and children behind, to send money home and support them. He's been in America for nearly a decade, and until today has never set foot in a courtroom. 'I'm not nervous about being here,' he tells me, even as he explains he's living illegally, in the shadows.

Cesar lives modestly, cutting grass and working in Virginia's expansive fields harvesting crops in the summer. He's never tried to get legal status. 'It's too difficult to get permission,' he says as he fiddles nervously with the baseball cap. 'Thank God I've never had any issues,' he adds, as a small smile flickers from under his moustache.

When Cesar crossed the border, things were easier. For Luis's generation the stakes are higher. He came to help his father out. More families are travelling to America, and it's created a backlog of cases in the already congested court system.

So it could take a long time for Luis to learn whether he can avoid deportation. President Obama is pledging to reform the country's immigration system to bring people like Cesar out of the shadows, allowing them to work legally and pay taxes. But with many states bitterly opposed to his plans, lengthy legal battles lie ahead.

'Bear with us, we have five thousand cases,' says Judge Bryant, as he calls the next person up to be heard. A girl named Diana heads to the front. Wearing a scarlet lace dress, her hair plaited on one side, she too carries a file of documents. It's hard to believe that at the age of fourteen, she is already sitting before a judge, her legs barely touching the floor as they dangle from the chair. Her hearing is also short – the judge tells her to find a lawyer and come back later in the year.

Diana, who's from Honduras, travelled by bus to Mexico with her sixteen-year-old sister. There they crossed the river to America in an inflatable boat. As she awaits her fate, she is in Virginia, living with her aunt. 'I'd like to be a teacher one day,' she tells me, grinning. 'Maths, I think,' she adds.

As I leave I peer into the adult courtroom, where a man dressed in a smart suit is finishing up his hearing. The man, who asked not to be named, came to the USA in 1995 from El Salvador and has raised four children here, but despite multiple hearings, has no legal status. 'Come back in 2018,' he's told by the judge. This court alone has nearly 20,000 cases pending – some of which won't be heard until November 2019.

It's estimated there are more than eleven million undocumented immigrants in the United States. But, for so many, the American Dream they came in search of has turned into a waiting game. One which could still be going on well after President Obama leaves office.

John Sopel – Trump's Super Tuesday Surprise – 05/03/16

Donald Trump's relationship with the press was combative long before his presidency was even imagined. But his campaign rhetoric about the 'lying press' and 'fake news' being churned out by 'enemies of the people' raised the temperature even further. At his rallies, reporters would be pointed out to the crowds – to single them out for jeers and disdain. Yet Trump's ability to wrong-foot his critics could never be underestimated. John Sopel was taken unawares by an unexpected show of charm.

Never let it be said that Donald J Trump is predictable. On the night of his greatest victory on this extraordinary journey, the playbook was obvious: you pack a giant ballroom with thousands of cheering supporters chanting U-S-A, U-S-A, holding Trump banners. It would be an excitable sea of red, white and blue – and the man of the moment, wooing the crowd in his inimitable style, bragging about his Super Tuesday heroics. At a stroke all the TV networks would have the desired images for the morning shows.

Instead, we sat huddled in the Mar-a-Lago resort, this chi-chi, one-time privately owned mansion in Florida, now a members' club owned by Mr Trump – and instead of supporters, he's chosen reporters. He's giving a news conference to a bunch of journalists who want to ask him impertinent questions.

Why would he want to do that? Well, having travelled with Donald Trump to California, Texas, Iowa, New Hampshire, South Carolina and now Florida, I can say the one thing you learn is: there are no rules. And the unexpected will always win out over the expected.

He doesn't so much give a speech as offer a series of disconnected thoughts, where he starts down one road, then distracts himself and goes down another. He can be talking about China and before the sentence has ended he's going on about Planned Parenthood and then about how rubbish his opponents are. But you come away clear about the themes.

He wants to make America great again, as he puts it. He wants to build a wall between the USA and Mexico – and it's going to be a great big beautiful wall and the Mexicans are going to pay for it. All Muslims will be kept out – though he's never explained how that would be enforced. And under him, America is going to start winning again.

He hates politicians and dismisses them as all talk and no

action. I remember, years ago, travelling every day with the British prime minister John Major, when the Conservative Party suffered a crushing defeat at the hands of Tony Blair. I can still recite whole passages of John Major's stump speech, I heard it so often. Donald Trump has stump *ideas:* it's an anagram of a speech, with words, ideas and bits of sentence, syntax and grammar having been put into the equivalent of the green Scrabble bag, shaken vigorously and thrown out onto the board.

We sat in what looks like a small anteroom of the Palace of Versailles – all white and gold, massive crystal chandeliers, Cupids in the cornices, delicately inlaid floors. And instead of Louis XIV with his powdered wigs, we await 2016's version of the Sun King, and his own equally exotic arrangement. Just think how much thicker the ozone layer would be were it not for all that hairspray.

And when he eventually appeared the unexpected happened – as we just knew it would. Donald Trump came over all conciliatory.

He was quietly spoken. He said he would be a unifier – of the Republican Party, of the nation. He didn't crow and he didn't claim to be the nominee, but he clearly thinks the primary race is effectively over. Welcome to Donald the magnanimous. He graciously congratulated Ted Cruz over his wins in Texas and Oklahoma. No mention of him being 'the biggest liar' he's ever met, which is his usual line. No demeaning of Marco Rubio either.

You might have been forgiven for thinking an impostor had entered the room. But no: it was Donald 2.0 we had with us. The trouble, though, when you upload a new operating system, is there are inevitably bugs and glitches. The new issue takes a bit of getting used to. And there will be many who say what brought them to the product was the original software.

So can and will the new magnanimous Donald be able to keep up this latest modus operandi, and will his army of fans like what they see?

The first two rows had been reserved for VIPs. They were coiffed, lacquered, expensively attired and perfumed – and that was just the men. The women were testament to the seductive power of plastic surgery. There were tucks, nips, lifts, implants galore. This was the

Trump rich crowd. And they seemed to like what they heard.

But what has made him a phenomenon is his ability to speak to the anger of blue-collar America. The ordinary Joes who think the American Dream, with social mobility at its core, has passed them by. People who worry that their children will grow up worse off than they are.

Last weekend I was in Georgia, and travelled about forty miles south of Atlanta to Senoia. They were dirt racing – poor man's Formula One. Big engines on cobbled-together car bodies going round a circuit not much bigger than a running track at phenomenal speed.

It is fast, furious, macho and noisy. Not unlike Mr Trump himself, one might say. And we were allowed to go into the central ring where the cars were being tuned and revved before going out to race. We spoke to mechanics and drivers, owners and hangers-on.

Normally, when you do interviews like this you get a range of views which you try to edit fairly. Here? No need. Everyone, and I mean *everyone*, was a Trump supporter. One told me he hated the Democrats and hated the Republicans, and loved Trump because he was the 'screw all of you' candidate.

And these are not just supporters. They view him as a saviour. A lifeline. A man who will bring back hope and opportunity. A man who will transform Washington politics and America's place in the world. He is their champion. A billionaire who grew up moneyed and privileged, the hero of the downtrodden. Just one of the many unlikely storylines in a totally improbable script in this most bizarre election cycle.

Linda Pressly – Opioids: A Death in the Suburbs – 27/08/16

As the saying goes, 'everybody's gotta die of something', but it seemed a grim milestone when drug overdoses became the leading cause of accidental death in the USA. They now kill more people each year than car accidents – and far more than gun violence. A particular factor is widespread long-term addiction to opioids – often powerful, widely prescribed pain medication as well as street heroin. Ohio, in the

Midwest, is one of the worst-affected states, and reports opioid deaths in rural, suburban and urban areas, at all social levels. Linda Pressly went on call with the coroner of Lorain County, in the upmarket suburbs west of the city of Cleveland.

The street could hardly be leafier. Water sprinklers turn lazily on lawns, the stars and stripes flutter from garden flag poles, and three small boys dressed in shorts pedal by in the hazy heat. We pull up close to Dr Craig Chapple's car. He's got the boot open, and is sorting out the equipment he'll need. Dr Chapple is no-nonsense, direct without being rude. When he talks, his white moustache – twisted and waxed at the ends – moves like a very small animal threatening to make a getaway.

The house the coroner is attending is a large bungalow, with a double garage. There's a rockery, pink petunias and yellow roses. Filigree lanterns and small, red, glass hearts hang from shrubs.

We wait at the end of the drive while Dr Chapple explains our presence to the family, and assures them we won't be identifying anyone.

Also milling around outside are the police. There are two coroners' assistants, eyes peeping out from under pulled-down baseball caps, their hands covered by surgical gloves. And there's the family. A man looks deeply upset. And we hear the coroner talking to a woman who's sitting at a garden table, sipping liquid the colour of rusty nails through a straw, holding a glass shaped like a small goldfish bowl.

We hear her tell the coroner she just found him like that, early this morning. Her brother was staying temporarily with her and her husband; last night they left him watching TV, and the next thing she knew . . .

Well . . . we're about to see for ourselves the next thing she knew. Dr Chapple calls us over, introduces us to the family, and we offer our condolences. We follow the coroner into the house.

In the large living room the television's on, tuned to the weather channel. It's really loud, but no one else seems to notice. There's a standing fan – it's on a high setting, and its whirr competes with the din of the TV. On the sofa, the woman's dead brother looks as though he's just sleeping. He's upright, his head rolled to one side, eyes closed, a rug loosely pulled over his knees. He's probably in his late thirties.

Dr Chapple gets to work. He investigates the scene, snapping away with a small digital camera. Then he instructs his assistants to remove the blanket covering the man's legs. They tug at it gingerly; 'Gotta watch out for needles,' says Dr Chapple. He examines the man's arms and legs to see if there are any puncture marks from injecting heroin. Nothing ... He tells his assistants to move the man onto the floor, and to search the sofa.

After a methodical examination of the living room with a tiny torch, Dr Chapple finds no physical evidence of opiate use. Of course, if someone's taken pills or snorted heroin, that might not be immediately obvious ... And the coroner says it's not unusual to find nothing – often the family clears the decks before calling the police, worried about the presence of illegal drugs, or shocked and shamed by a loved one's overdose.

Dr Chapple calls for the mobile stretcher. He shakes his head. This is a too-familiar scene for him – on one weekend alone, he says, he attended four drug deaths. It's shocking. Lorain County has a population of just 300,000. It's mostly suburban, middle-class, with a large rural hinterland. But for the last three years the number of fatal opiate overdoses has hovered around the sixty-five mark. But the same number of deaths has been recorded in the first half of 2016 alone. Mainly this is a reflection of the unleashing of a far more lethal assassin. Fentanyl is a synthetic opiate many times stronger than heroin. But drug dealers are mixing it with heroin; either the addicts think they can take this powerful cocktail, or they're not aware what's in their fix.

I step into the hall while the coroner finishes his work. Although the TV and fan remain on, I still hear the crackle of industrial-strength plastic, and the zipping up of the body bag. I chat to one of the police detectives – it's the second suspected overdose he's attended this week. There's a lull in our conversation as the body's wheeled past. And then he says, 'So ... what about Brexit?' For a moment I have a profound feeling of giddy disorientation. His question's so incongruous, I almost laugh. But this swift and unlikely change of subject is indicative of how drug overdoses have become just another regular part of his day job.

Aleem Maqbool – Confederate Monuments – 03/06/17

'The past is not dead: it isn't even past.' As a native of Mississippi, author William Faulkner had a finely attuned sense of the South's intense attachment to history. That history is still passionately fought over today, as people summon the symbols and the rhetoric of earlier centuries to invoke their own visions of what America should be. As with South Africa's #RhodesMustFall movement, the attention of those campaigning for more equality in the present day has been turned to statues of historic figures. Aleem Maqbool reflected on how important the stories we tell can be.

At two o'clock in the morning, at the base of a statue I'll admit I'd never really noticed on my many trips to New Orleans, a group of a dozen or so protesters had gathered . . . many draped in, or waving, Confederate flags. Young people were shouting abuse at them from a nearby bar – some even came across to confront them.

It was a statue of Jefferson Davis. He was the Confederate president during the Civil War and a defiant defender of slavery – and parks, roads and schools are named after him all over the southern part of this country. After many years of debate, and legal process, the city council was going to take down the monument.

Those draped in flags came out every night to try to stop that happening. 'They're destroying a piece of Southern culture,' one man told me. Another protester insisted, 'These people were good people. They lived and died for a reason. Why should we just forget it?'

But a few nights later I watched a crane lift Jefferson Davis up off his pedestal. His body gently spinning around, his outstretched arm sweeping from left to right, he disappeared into the back of a truck.

In covering America's race problems, I'm always asking myself what the solution might be. Chanting 'Black Lives Matter' is one thing, but what tangible step forward could demonstrators demand to bring about racial equality here?

A couple of years ago, in Alabama – as I worked on pieces to mark the fiftieth anniversary of Martin Luther King's march from Selma to Montgomery – I asked one of the most respected African-American civil rights voices of today, lawyer Bryan Stevenson.

His one-word answer surprised me: 'Monuments.' Not police body-cameras, not a specific change in the criminal justice or education system ... A man who'd thought about this question for decades told me that 'monuments' were what America needed to address to move forward on race. 'In America, we have monuments and markers all over about the Confederacy,' he said. 'We have very few that talk about slavery or the terror era.'

The more I travelled, the more I saw how right he was. But how was changing that going to address the terrible racial disparities that I'd also seen?

I've reported from Ferguson, where the unarmed black teenager Michael Brown was shot dead by a police officer and his body left on the street for hours in the summer heat. I'd seen the confrontations between police and Black Lives Matter protesters there ... and in Baltimore and South Carolina and many other places. Would replacing statues really help stop that?

But they are talking about more than statues. They are also talking about 'narrative'. Bringing down Confederate statues across the South is the obvious bit. Many were put up specifically to taunt African Americans after slavery was abolished.

There are very few monuments to mark the sites of old slave markets, the ports where the shackled men and women were brought ... and, from later decades, nearly 4,000 sites where black people were lynched go entirely unmarked.

It's about even more than that. In many places in the USA, schools can teach history however they want – so some students learn about the Civil War without even a mention of its being a fight over whether or not slavery was abolished. But why does it matter?

While in New Orleans to cover the removal of monuments, I was working with a German colleague called Franz. We've talked a lot about how he was taught about the Holocaust at school. He remembers going to a football match aged fifteen and wondering whether he should even be proud to sing the national anthem, so ashamed was he of what he'd learned.

Germany was only able to move forward when it was brutally honest about its past. That is an issue for the whole of the USA, not just the southern states.

Canada

Sian Griffiths – Aboriginal TRC – 11/06/15

For over a century, as part of a forced assimilation policy, 150,000 children from Canada's indigenous peoples – often called the 'First Nations' today – were sent to church-run boarding schools. Seven generations of children were separated from their families and stripped of their language and culture. Many faced emotional, physical and sexual abuse. Thousands died of neglect and disease. The last such Indian Residential School finally closed in the 1990s.

Canada decided to appoint a Truth and Reconciliation Commission to assess the impact of this system. In 2008, the Canadian government apologised for its actions, acknowledging the policy had left many Native people – and their communities – broken and dysfunctional. Sian Griffiths was present as the chair of the commission released his findings to an emotional audience.

I wriggled my way through an expectant crowd in a ballroom of a downtown Ottawa hotel. There was standing room only. Several hundred people, among them former residential school students, known as 'survivors' – many of them now elderly – stood shoulder to shoulder with camera crews. The event was going to be broadcast and streamed live across the country. Rarely had an aboriginal-led event attracted such national attention. Everyone was waiting eagerly for the long-awaited findings of the Truth and Reconciliation Commission.

While this moment was the culmination of a seven-year enquiry, many survivors had waited decades to share their stories. Many had felt such shame at the abuse they had endured that they never told another living soul – not their families, not their friends – until the commission arrived in their corner of Canada. The commission would allow them to at least begin their healing journeys.

The enquiry was more of an endless marathon, an odyssey. The three commissioners travelled tens of thousands of miles to hundreds of communities, even the most remote on the shores of the Arctic Ocean, to hear story after story of abuse. By all accounts, the work of the commissioners was relentless, physically demanding and emotionally gruelling. With the evidence now gathered, Canada was about to face a moment of reckoning.

The room fell silent when Calvin Murray Sinclair, the chair of the commission and Canada's leading aboriginal judge, approached the podium. Tall, commanding and august, his white hair seemed to have gone even whiter under the weight of the many sad stories he now carried with him. As he introduced himself, the room burst into a thunderous, deafening applause, accentuated by energetic drumming, wolf-whistling and cheering. It was the kind of welcome normally reserved for a hero returning from a difficult mission.

Justice Murray Sinclair, who described the residential-school era as one of the 'darkest, most troubling chapters' in Canadian history, began to read out his findings – more of a list of charges. It was clear this was going to be a prosecution.

He said that seven generations of children were 'stripped of the love of their families, their self-respect and ... identity' over the course of a century. In a deep, booming voice, he accused Canadian governments and churches of trying to 'erase from the face of the earth, the culture and history of many great and proud peoples'. He said there had been 'discrimination, deprivation and all manner of physical, sexual, emotional and mental abuse'.

His words obviously triggered painful childhood memories. All around me, people began sniffling, shedding tears and even sobbing. Counsellors handed out tissues and glasses of water. It seemed that people hadn't just gathered to hear the report but to experience a collective emotional release.

Then, with two words, he issued his damning verdict: 'Cultural genocide.'

Again, the room erupted into massive applause. For them, it was a long-awaited affirmation. But for many Canadians this would be bewildering news. Weren't they, after all, the nature-loving, nice guys? Justice Murray Sinclair was showing them their dark side.

He left it to a fellow commissioner to reveal the tragic news that 6,000 aboriginal children had died of malnutrition and disease at residential schools. Those were just the ones they knew about. Many records, she said, had been destroyed. The school authorities clearly had little regard for the families; many were never informed about the loss of their sons and daughters. Often the boys and girls were buried on school grounds, which she described as having 'cemeteries but not playgrounds'. 'We can't un-know what we have learned,' she said.

Justice Murray Sinclair blamed the residential-school system for the dysfunction, chaos and poverty in aboriginal communities today. They face high crime, addiction and unemployment rates, and poorer than average prospects in health and education.

With the truth out, now came the reconciliation. Justice Murray Sinclair laid down ninety-four recommendations, really preconditions to true reconciliation. Though one of them called for a formal apology from the Pope – many of the schools were Catholic-run – most urged Canada's political leaders to commit to closing the gap between aboriginal and non-aboriginal Canadians. Thus far, the government has agreed to just one. However, in a sure sign that an election is looming, opposition parties agreed, if elected, to implement all the recommendations.

One of those key demands is that this shameful chapter of Canadian history be taught as a mandatory subject in all Canadian schools. One key date students will never forget will be the day Justice Murray Sinclair made Canadians face up to their past.

Rajini Vaidyanathan – Trudeaumania! – 22/10/15

European and US citizens sometimes wonder why their government can't be more like Canada's. It seems a land of integrity, moderation and quiet prosperity – as well as a compassionate and constructive voice in world affairs. During the nine years of Conservative rule under Stephen Harper, that image took a bit of a bashing; it turned out that Canadian voters, too, could be swayed by populism, tax-cutting and big oil profits. The election of Justin Trudeau, son of the late prime minister Pierre Trudeau, in 2015 marked the end of Harper's era and triggered much praise of the new man's charisma, poise and all-round niceness. Rajini Vaidyanathan wondered how far image might have triumphed over substance.

The night after the election, as morning broke in downtown Toronto, I met Chris, a fast-food worker in the city. Wearing baggy grey tracksuit bottoms and bright red trainers, his cap was embroidered with the country's symbol, the maple leaf.

It would've been hard to miss him as he bounced along the street. His scarlet trainers moved up and down in a steady rhythm as he chanted, 'Trudeau, Trudeau,' occasionally punching a fist in the air.

Hours earlier his Liberal Party had pulled off a stunning victory. 'I'm excited,' said Chris. 'I think Justin Trudeau is going to do a great job for this country ... I think Stephen Harper did nothing but make our country horrible,' he added.

Chris, like many others I'd met through the course of the campaign, wanted change. And 'Real Change' had been Justin Trudeau's campaign slogan.

Days before the election, I found myself in a small Indian snack shop – the smell of chai and samosas making my stomach grumble – as I waited, along with several dozen members of Toronto's Sikh community, for the Liberal Party leader to make a campaign appearance.

His enlarged face beamed from a campaign advert on the side of his election bus as it pulled in. In one shot he was posing in a fur-trimmed coat with husky dogs in the snow, in another he was at ease in a blue-collared shirt as he sat with a young family.

When the man himself stepped down from the bus, he was sporting yet another look, wearing a blue sports jacket with 'Canada' written on the back. He looked more preppy than politician. But once inside the restaurant, the jacket was removed to reveal a crisp white shirt and a spotty purple tie.

'Why don't you take a selfie?' Trudeau suggested when one eager supporter asked him for a photo. He leaned in as the pair smiled together, before moving on to the next person. Justin Trudeau is the politician for the smartphone generation – young, trendy and media savvy.

During the campaign, his opponents said he wasn't ready for the top job, arguing he lacked legislative experience because he'd only been an MP since 2008. They poked fun at his good looks, saying he was nothing more than a pretty boy, making several references to his well-groomed hair. But Mr Trudeau turned this apparent disadvantage around – making a virtue of his fresh approach to politics, and his youth.

In the end his Liberals took seats from both rivals, the Conservatives and the NDP, the New Democratic Party. And with nearly 40 per cent of the popular vote, and 184 parliamentary seats, Justin Trudeau's mandate is decisive. But that still leaves a large percentage of Canadians who didn't support him, unconvinced.

'I didn't vote for him. I picked the devil I know, the Conservatives,' said Tim, an engineer I met as he sucked slowly on a cigarette during a break from work. 'It's not surprising he won. Trudeau caught their attention,' he said, referring to the large number of young voters who turned out in this election. 'He's closer to their age group,' he added, taking another puff.

Close by I also met Shauna, who at fifty-two is nine years older than Justin Trudeau. 'I didn't vote Liberal,' she explained. Shauna comes from a Liberal family, who adored Justin Trudeau's late father Pierre Trudeau, who served as Canada's premier for more than a decade. Shauna remembers seeing Trudeau Junior grow up before the country's eyes, but that familiarity wasn't enough for her to cast her ballot for him. 'Part of me thinks it's cool that he's following in his father's footsteps, but his platform is so different – he has a lack of fiscal understanding,' she argued.

Justin Trudeau campaigned on a distinctive policy platform – investment and spending over austerity, a plan to withdraw Canadian forces from airstrikes against IS, and proposals to legalise marijuana. For many like Shauna, he's untested, and presents a risk. Others, like Chris, see him as a breath of fresh air. As prime minister, Justin Trudeau will have to work to reconcile some of these differences if he wants to build an inclusive government.

Shauna may not have voted for him, but she did have one view of Mr Trudeau which many seem to share. I heard it from several people on the campaign trail. It was as if they were sharing a guilty secret. 'I think he's really handsome ... No, seriously!' she told me, laughing somewhat uncontrollably.

Trudeaumania may be back in Canada, but the country's new prime minister must surely know: it's not just about how things look; it's about whether you can deliver.

South America

Will Grant – Venezuela without Chávez – 16/03/2013

Where to look when an intensely charismatic leader steps away from the spotlight? Returning to Venezuela as Hugo Chávez's health worsened, Will Grant found the country on tenterhooks. The Comandante had made an unforgettable impression on the world stage, in the vanguard of a so-called 'Pink Wave' of left-leaning governments across South America – and at this point his ruling PSUV party also commanded great loyalty at home.

Caracas is unlike any other city in Latin America. It's hard to explain the chaotic, concrete mayhem that is the capital of this oil-rich nation, how its gridlocked traffic sits beneath an untouched, lush green national park. It's difficult to accurately portray its rampant inflation and nonsensical black-market currency or its constant, brutal urban violence.

Or the polarising effect of President Chávez. Above all else, for the past fourteen years, Venezuela has been characterised by Hugo Chávez.

As we made our way through the airport his was the first face I saw on arriving back on Venezuelan soil, just as it had been the last image I'd seen when I left my posting in Venezuela in December 2010. A huge poster of 'El Comandante' overlooks the arrivals lounge, proclaiming that 'The revolution is moving forward'. Yet something was different. The picture of Mr Chávez was quite recent: the toothy grin was the same, the clenched left fist raised aloft. But he now looked bloated in the jowls and cheeks, perhaps a consequence of the repeated rounds of chemotherapy he's been through in Cuba.

Suddenly it struck me as we drove up into Caracas: this would be the first time I'd experience Venezuela without Hugo Chávez. I first arrived in 2007 to help cover the controversial switch-off of an opposition television channel; I ended up staying for twenty-six months, in the kind of love-hate relationship you can only experience in cities and countries which truly take you somewhere new. Caracas, and Venezuela, had certainly done that.

Chávez had been the constant during those years. The omnipotent power at the top of the Bolivarian Revolution, deciding its direction, shaping its future, controlling its destiny. Famed for his hours-long speeches, he could whip up the faithful into a fervour, offering hope to his supporters while enraging his opponents. Whatever you thought of Mr Chávez, you simply couldn't ignore him.

And now, suddenly he wasn't here. I'd been in Cuba when he arrived for the operation from which he's still recovering. That was two weeks before Christmas and he hasn't been publicly seen or heard from since.

This is a man who loves the limelight, and there's no way he'd miss his inauguration unless he was simply too ill to attend. It seems his formidable stamina is being tested to its limits by the cancer. But we don't know for sure how ill he is, nor even what kind of cancer he's suffering from. The Cuban medical team treating him is well-versed in keeping its patient's health a state secret.

Mr Chávez has worked for fourteen years without a break, his supporters say. One of them, Leni Beatriz, has been trying to rationalise his absence. 'I'm imagining that he's on holiday, taking all that time off that he's owed,' she said as she took us around the streets of her staunchly pro-Chávez neighbourhood, San Agustín. 'We'll be here, supporting him and waiting for him to get better. We, the Chavistas, the revolutionaries, are more united now than ever.'

But he isn't on holiday. He hasn't sent a message of support nor thanks to his concerned followers back home. And no one who spoke on the day of his inauguration dared to venture when he would be coming back to Caracas, or if he was coming back, nor whether he'd be able to return to work even if he did walk through the doors of the presidential palace, Miraflores.

The uncertainty is unsettling for both his friends and his foes. Neither side has a handle on what will happen next, and the opposition are at sea over how best to tackle this unprecedented situation.

Still, Chávez's followers remain committed. I met Javier García, who says he will vote for Chávez for as long as he lives. And should Mr Chávez go first, Javier will vote for his successor. Little wonder when you consider how his life has changed since the socialist leader took office. Once a drug addict living on the streets of a shanty town, he got clean through a state-run social programme, now has a government job and a shiny new apartment, all courtesy of the Bolivarian Revolution.

His cropped hair and broken front tooth are all that remain of the thug life behind him. Javier began to weep when I asked him about life without Hugo Chávez. 'The revolution has to be like this building,' he said, wiping away the tears with his sleeve. 'Earthquake-proof. This is a strong tremor, but it won't fall down.'

Vladimir Hernandez – Venezuela – A Bitter Cup to Swallow – 28/02/18

Adding up the true bill run up by government policies was never easy in Venezuela, as it seemed to defy economic gravity for so long. With spiralling inflation and perennial shortages, Venezuelans had to adapt to survive – though when President Nicolas Maduro suggested in 2017 that people should breed rabbits and grow vegetables on their balconies to boost food production, he was widely mocked. Making his own return visit to the country of his birth, Vladimir Hernandez was brought up short by one graphic example of economic surrealism.

The coffee of choice for many Venezuelans is a kind of double macchiato. Called *marron,* the word for brown, it provides a sharp intake of caffeine, often early in the morning or after a hefty lunch.

As I sat at a restaurant table in Caracas, I almost choked on mine when the bill arrived. The delicious coffee I had just downed cost the same as I had paid for a flat here fifteen years ago. I still remember how the mortgage payments for my little one-bed place would gobble up most of my income, as well as the enormous sense of achievement in buying my first property after growing up in a working-class family. So as I looked at the empty plastic cup in front of me, fifteen years of my life flashed rapidly in front of me. Welcome to Venezuela's world of hyperinflation.

The Venezuelan economy has collapsed since 2014. The private sector has shrunk and the state has resorted to printing money to try to sustain the huge spending that has led to the current crisis. Simply printing more and more money is a formula that, many economists will tell you, leads to disaster. And disaster seems to be on the cards for Venezuela: the IMF estimates the inflation rate could reach 13,000 per cent by the end of this year – a mind-boggling number. Today a kilo of onions costs almost as much as the monthly minimum wage. I know someone who needs drugs for their high blood pressure; each box costs her a third of what she makes in a year.

I've been reporting on the crisis for three years now, travelling regularly from London to Venezuela. I've seen children losing their skin because of terrible malnutrition. I've seen an eight-year-old who looked no more than three, his growth stunted due to a lack of food. And I've seen people crying . . . lots of them. On my last trip, four interviewees in a row broke down in tears after they told me how they're struggling to find food. One was an unsettlingly thin mother, her child in her arms, who doesn't know what to say to her children when they're hungry.

I've also seen how much weight my own friends and relatives have lost. This, for me, is not a story that goes away when I leave the country. My brother – who lives there – sent me a message recently. He had lost his job and was struggling. He told me that because he lives near a big lake, he's now resorted to fishing to get food for his two young daughters.

How did it come to this? I kept asking myself this over and over again. How did this country, with the biggest oil reserves in the world, reach a point where it cannot feed its own people?

I asked one of the new faces in the ruling party this exact question. His view – and that of all government ministers – is that the USA and its allies are to blame for imposing sanctions on the country last year, accusing Venezuela of collapsing into dictatorship.

However, this is not a crisis that started last year. The government has been peddling another line too, calling the economic collapse the result of an 'economic war' waged by right-wing factions inside and outside Venezuela. In this version of events, those factions are seeking to topple President Nicolas Maduro and frustrate the march of the socialist Bolivarian Revolution which reached power almost twenty years ago.

As he was retelling me this official line, behind him I noticed the face of the young woman who had been cleaning the room as we walked in. She was small, thin and incredibly cheerful – as many Venezuelans are – despite the fact that she'd told me she eats just twice a day.

As she heard the official, I watched her expression turn sour. I couldn't help thinking that this low-wage worker was once part of the huge wave of enthusiasm that brought the socialist government of Hugo Chávez into power. And it was people like her, the poor, who have kept Maduro in office. The look of disgust on her face made me think that this is a revolution that may now have left the poor behind.

Katy Watson – Bolivia – El Alto Rising – 11/10/14

Hugo Chávez wasn't the only populist left-wing leader to ride a wave of support from the downtrodden during the early 2000s. In Bolivia, Evo Morales became the country's first leader of Aymara descent. After rising from a humble background to prominence as a leader of rural coca growers, he won three terms as president – and still angled for more. Around three dozen indigenous groups make up nearly half of the Bolivian population, while the vast majority nationally have some indigenous heritage. So Morales's message of improved opportunities

*for his people and greater pride in that heritage found a ready audience.
In La Paz in 2009, Katy Watson sensed optimism in the air.*

I first visited Bolivia when I was a young student of Latin American
politics. Coming overland from Chile by bus, arriving in La Paz
took my breath away. High up in the Andes, La Paz is a city that
feels like it should never really have been built – it's so impractical
and remote. Nestled in a canyon below snow-capped mountains,
houses are squeezed onto every last bit of available land, rising
precariously up the side of the cliffs.

On this last trip, La Paz took my breath away again – but not in
quite the same way. I flew in this time and as I was waiting for my
luggage I spotted the free oxygen on offer to passengers. I chuckled
– only in La Paz, I thought. On my first night, I went out to meet
a friend. I had a glass of wine and thought nothing of it. Only the
next day I could think of nothing else. Altitude sickness, or *soroche*
as they call it here, had hit and I was bedridden with a stinking
headache and nausea all day.

Twenty-four hours later, I was acclimatised and fit to hit the
streets. The roads are still potholed and the traffic terrible. But
since my last visit, there's been an addition to La Paz's incredible
skyline – a cable car. It's a big deal. When it opened earlier this
year, people queued for hours. Rather than squeezing into cramped
buses, riding this space-age glass bubble, complete with Wi-Fi, is
just out of this world.

The cable car's main route is between La Paz and El Alto, which
sits above the capital on the high plains of the Andes. More than
just a symbol of development, it links two cities that are often seen
as very different. El Alto used to be a suburb of La Paz but, as it
grew, became a city in its own right. Its residents are migrants –
people who've come from the country in search of work – and it's
now growing at a faster rate than richer La Paz.

Altenos, as the residents are called, are mainly indigenous Aymara
and some Quechua. Most women here wear their traditional dress –
long black plaits, bowler hats and colourful skirts known as *polleras*.
And they carry an *aguayo* on their backs – a fabric sling that's used

for everything from babies to the weekly shop.

El Alto has long been seen as La Paz's poor cousin but there's hidden wealth here that is slowly starting to reveal itself.

Freddy Mamani is at the forefront of this change. He's from El Alto but studied architecture in La Paz. He has become the king of Andean architecture, building dozens of houses that are changing the face of the neighbourhood which has, up until now, been full of simple red-brick and concrete houses.

He took me to one house under construction. As well as private accommodation, the first and second floors were taken up with a dance hall. I have to say I've never seen anything quite like it. Bright orange and yellow paint. Ceilings fifteen metres high with space for an orchestra and an upper balcony for guests. This was no village hall entertainment. I felt like I was in Disneyland.

Freddy's style, he says, is influenced by Tiwanaku, one of the most famous archaeological sites near Lake Titicaca. It's known for its incredibly accurate stonemasonry. Of course, Freddie has a modern take on it. Using bright coloured paint, geometric patterns and folkloric elements, he says he's broken with traditional architectural rules.

But not everyone appreciates his designs. People have started to call them *cholets* – a combination of 'chalet' and *cholo* or *cholita* – which is a slang word widely used to refer to indigenous people here. Calling somebody a *cholita* may now be an affectionate term, but Freddy doesn't like the label. *Cholet*, he says, is pretty demeaning.

Freddy took me to meet another client: Alejandro, an Aymaran businessman who's worked as a tailor since the age of fourteen. He saved up more than half a million dollars to pay for his new palace in cash. The result? Not just one dance hall but two, each with a capacity of a thousand people. He rents them out every weekend; the chandeliers were imported from China, he tells me. The walls were hand-painted with designs showing off Aymaran culture and lights flashed from every luminous-green pillar. But Alejandro wasn't finished. I followed him up some stairs – and what do you find on the fourth floor? A small football pitch of course!

I asked Freddy where this new wealth was coming from. Have

Evo Morales's indigenous-friendly policies made people richer? 'They've always been rich,' he told me. 'What's happening now is that people feel prouder of their identity – they want to show it off.'

Drug Wars

Robin Lustig – Peru – Coca Changeover – 30/11/13

The campaign against coca growing in Colombia's countryside demon-strated the notorious 'balloon effect' – if drug production fell in one region, it was likely to swell in another before long. Estimates of the amount of land under coca cultivation in Peru soon started to rise. Robin Lustig made his own Andean journey to see what was happening on the ground – perhaps without realising at first how demanding it would be.

I should probably have listened a little more carefully when the farmer answered my question. I'd asked if she would show me where her hidden coca plantation was, and what she said was, 'Yes, of course, but it'll mean a bit of walking.'

Now I like walking; I walk for pleasure. But what a Peruvian farmer means by a 'bit of walking' turned out to be rather different from what I mean. We were in the region known as the High Amazon: green, lush hillsides and steep wooded valleys, where the foothills of the Andes meet the Amazon jungle.

Traditionally it's been one of the main production centres for Peruvian cocaine, and although the government is committed to eradicating the cultivation of the coca leaf from which cocaine is made, there's still plenty of it about.

Hence my walk in the jungle. We set off, following the farmer, her husband and their dog, down a steep muddy path. He carried a machete, and hacked away at the undergrowth to clear a way for us down the side of the valley. After I'd already landed on my backside half a dozen times, he cut a stout pole to help me keep my balance.

It took forty minutes or so of slipping and sliding before we reached the bottom. And, as is often the case at the bottom of valleys, there was a river. Not a huge river, admittedly, but still a river, and we obviously were going to have to cross it. There wasn't a bridge, but there was a fallen tree trunk. By this time, I was carrying two stout poles, plus a backpack, but I was helped across by the farmer's husband, who gripped my hand tightly as I inched along the tree trunk, and their dog, splashing excitedly through the water, tail wagging, egging me on.

Clambering up the other side was a lot easier than sliding down had been. When we eventually reached the coca clearing, I finally got a chance to catch my breath and to talk to our companions. I never learnt their names – they thought it was probably better for all of us if they remained anonymous. After all, as far as the authorities were concerned, they'd given up their illegal coca cultivation and were now growing coffee and bananas. Which, to be fair, they were – but as well as the coca, not instead of it.

So we sat and talked. She was wearing a pink T-shirt, a baseball cap, faded brown trousers and gumboots. Stockily built, in her forties, and defensive about continuing to harvest a crop that her government says she shouldn't be growing. 'We can't make a living growing just coffee and bananas,' she insisted. 'It takes years for those trees to become established, so what should we live on? We get no help from the government, so we have to grow the coca.'

She would love to stop, she said – but that may have been because she knew that was what she was meant to say. Yes, she knows what happens to the coca leaves after she's sold them, and she knows that criminal gangs control the trade in cocaine. 'I'm only growing a very small amount,' she said. 'It really doesn't make any difference.'

The coca growers of Peru insist that they aren't the ones who get rich from the cocaine trade. Although I did meet one man – a former coca grower who has now switched entirely to cultivating cocoa instead of coca – who admitted that in his former life he did make a very good living. He spent his ill-gotten gains, he told me, on gambling and drinking and women. Now, as a solid law-abiding citizen, with three jail terms behind him, he says he spends his more modest income on looking after his family. 'I don't have as much money as I used to have,' he said. 'But I do sleep better at night. I don't have to worry any more about the police knocking on my door.'

Coca has been grown in Peru for thousands of years. If you chew the leaves, or use them to make tea with, they act as a mild stimulant, a bit like a strong cup of coffee. I bought a small bag of coca leaves, quite openly, on a street corner in the central Peruvian town of Tingo Maria. It's a thriving, bustling place, its streets jammed with Chinese-made motorcycle taxis – and it owes its prosperity, almost entirely, to coca.

They are trying to adapt. In a smart, New York-style coffee bar, I was offered a very good cup of espresso, made with locally grown beans. The hope is that it's caffeine, not cocaine, on which the region's future prosperity will be built.

Arturo Wallace – Colombia – The Struggle Is Not Over – 18/10/12

There are many ways to reach a story – from exhausting treks to alarming armed convoys – but exploring the terrain at first hand gives added insight. During its decades of political violence, many of Colombia's cities became ever more isolated from its rural reality. Bombings, kidnappings and armed assaults made moving around in certain areas extremely risky for outsiders – and turned travelling for pleasure into a thing of the past even for Colombians. The relationships between guerrilla factions, paramilitaries and drug traffickers were often so tangled they could be hard to pick apart. Arturo Wallace ventured into a former hotbed of the Fuerzas Armadas Revolucionarias de Colombia (FARC) at a point when the peace process between the government and armed groups seemed to be making progress, and some regions were opening up again.

The slope is so craggy that I feel like getting off my mule . . . but maybe that's not such a good idea. After all, she's the one who knows this trail, which goes deep inside the Cauca highlands, heartland of the Colombian armed conflict.

We have been trudging on for hours: my old, small and stubborn mule moving slowly up and down the mountains, stopping at will to nibble on the few patches of grass she can find next to the dirt track; me, marvelling at the landscape of steep green hills and deep canyons that disappear into the horizon.

This south-western region has long been a key corridor for drug trafficking, and one of the strongholds of the FARC, the left-wing rebel group that has been fighting the Colombian state for forty-eight years. This rugged geography is one of the reasons they have been able to carry on their fight for such a long time.

It was also here that FARC's main leader, Alfonso Cano, was killed last November. And on the hill where he was hiding, before the Colombian army attacked, I find some clues to what might have driven the rebels back to the negotiating table.

Very little remains of the small wooden house Cano used as a hideout. The craters left by the heavy bombing that announced the arrival of airborne commandos can still be seen, amid the young coffee trees now growing in the area. The local peasants who brought me here still shrug when they remember the sound of the planes and helicopters involved in the operation.

Over the last ten years, aircraft like this have given the Colombian army a clear military advantage over the insurgents. Many believe it's the air strikes which finally tipped the balance. The Colombian government is so convinced of this that the start of the talks has not resulted in a ceasefire. Military operations against FARC will continue until there is a final peace agreement. The fighting on the ground is *not* over – not yet.

On my way here, I came across 400 peasants – men, women and children – who had just left their houses after being caught in a gun battle between FARC and the army. They expected to be able to return home soon, but their worried and weary faces were a reminder that while Colombia talks about peace, there are still plenty of places where the noise of war sounds louder. 'You never really get used to it,' a local peasant of indigenous descent would tell me later. 'But we cannot abandon our land. How could we, if she is the one who provides for us?'

The fighting around this area has intensified over the last few months. But air power and airborne commandos are not enough to guarantee permanent control of these mountains. This is still guerrilla territory. FARC know we're here. They've been watching us – me and the International Committee of the Red Cross mission I'm tagging along with, as they travel the region to tend to horses and mules belonging to the local peasants.

It is not typical humanitarian aid, but remote rural communities cannot do without those animals, which are their main connection with the modern world – with health centres, hospitals and markets. Another important piece of the Colombian puzzle. For this magnificent isolation is, at the same time, cause and consequence of such a long-standing conflict.

Isolation protects the rebels, but also sustains them. For some local youngsters, becoming a guerrilla is the only job available. The lack of proper roads makes it very difficult to take products to market. Among the coffee trees and the sugarcane growing on these hills, every now and then I also see coca. The forbidden plant can be the only way to earn hard cash in such remote areas. But coca has also been the fuel that has allowed the Colombian conflict to burn for almost half a century.

From my mule it is evident why the Colombian government has accepted that military force alone cannot end the fighting. But as my four-legged companion keeps on trudging through Cauca, another thing also seems certain: if it really wants peace, Colombia can't stop at disarming the rebels. It will also have to create real opportunities for peasants – like the people growing coffee on the hill where the army killed Alfonso Cano.

Mathew Charles – Colombia – Rebels Still? – 27/01/18

Eventually the Colombian state and FARC did sign up to a much-celebrated peace agreement – but not all of the country's rebels were on board. Another left-wing guerrilla group, the Ejercito de Liberacion Nacional (ELN), also signed up to a ceasefire, but the agreement lapsed and peace talks broke down after the rebels attacked an oil pipeline and launched grenades at a naval base. Since it was founded in the 1960s, the ELN has repeatedly targeted large landholders and multinational companies, as well as oil infrastructure, claiming to be acting in defence of the rural poor. Mathew Charles explored the world view in its ranks and wondered about the long-term prospects for peace.

La lucha means 'the struggle' or 'the fight' in Spanish. It's a common answer in Colombia when you ask people how they are or what they're up to. For some, it's a *lucha* against the daily grind; for others, it's the real battle to survive. But for guerrilla fighters, it's why they joined up.

In the ferocious humidity of the jungle, I've been given free rein to roam around the makeshift camp. Some of the insurgents are playing football, others dominoes, and one after another they dutifully explain that their *lucha* is to combat capitalism, eradicate poverty and resist US foreign policy. Their answers seem automatic: monosyllabic and usually given without eye contact. But their *lucha* embodies a deep-rooted anger.

The shrill of a whistle calls the insurgents for inspection. They line up in military formation, but without much precision. The lines are not straight, postures are relaxed and some are not even wearing their boots. They begin to chant: 'Liberation or death! Liberation or death!'

Estasio has chosen not to join in. The inspection is optional, it seems. The guerrillas are his family, he tells me. They fed him and taught him to read. But he says the most important thing he's learned is class consciousness. 'I know who's out to get me, who exploits me,' he says. 'It's the rich.'

I'm struck by the age difference between the rank and file, who are mostly in their teens, and their articulate middle-aged commanders, identified by their red berets. The ELN denies that it has a hierarchical structure, but it is clear that the big boss of its Western War Front is Uriel. He meets me with a firm handshake and a broad flashing smile. He's tall and thin, in his early forties. His hair is black, though his beard is beginning to grey. He has a deep voice with the soft and distinct soothing melody of a *paisa* – someone from the Medellín region of Colombia. I'm distracted by his unusually long eyelashes.

Uriel is not wearing military camouflage like everyone else, but is dressed all in black. He wears the insurgents' distinctive armband with pride. It is black and red, marked with the letters 'ELN' in white. His rifle hangs awkwardly from his shoulder. He doesn't strike me as someone who's overly comfortable with a weapon, but I'm convinced he knows how to use it. His piercing brown eyes seem battle-hardened. His hands, in contrast, are perfectly manicured.

In apocryphal Biblical texts, the archangel Uriel is said to represent wisdom. The commander's choice of alias is a reminder of the ELN's religious origins. Unlike their more widely known counterparts, the FARC, the ELN's founders were not marginalised and impoverished farmers but radical priests and leftist intellectuals, inspired by the Cuban Revolution.

Uriel says the Colombian government is seeking a neoliberal peace, which he rather poetically defines as 'the silencing of weapons'. But for him, an end to the violence is not enough in itself. 'Our *lucha* is socialism,' he says, in another well-rehearsed answer. 'We want a change of model. If the government's objective is to demobilise us, we're wasting our time. This armed conflict can only be stopped if you attack its roots, which lie in poverty.'

Commander Uriel's protestations about socialism seem at odds with the designer labels visible in a pile of laundry behind him. The guerrillas receive no wages, I'm told, just the odd bit of pocket money. I ask how much the movement earns from moving its *mercancia*, or merchandise – the polite way of referring to cocaine. Uriel hesitates. He wasn't expecting this line of questioning. He rejects any criticism that the ELN lacks legitimacy, and grows more animated. For the first time, I sense his answers are genuine and not part of a pre-prepared script.

He admits that the guerrillas charge the drug traffickers a 15 per cent tax on each shipment moved through their territory. But he denies the charges made by the authorities – with some evidence – that the ELN has started to move the *mercancia* itself.

Determined to stay on-message, Uriel steers the conversation back to the negotiations with the government. 'We're in a waiting room,' he says. 'We want to see what the government comes up with. The ball's in their court.' This sort of evasiveness is why many Colombians doubt the ELN's commitment to peace.

There's another handshake as I get up to leave, though it's not as firm as the first. The smile's gone too. This is the *lucha* laid bare: desperate, fierce and contradictory.

Stephen Gibbs – Mexico – Vigilante Mayor – 03/12/09

In Mexico, too, the state has gone to war with the drugs cartels – which, just as in Colombia, have worked hard not just to make billions, but to use them to buy friends in high places. Even without the guerrilla armies of the Andes, the violence of the drug trade rippled through Mexican society. Abductions, ransom demands and intimidation spread everywhere. As state forces targeted the cartels, criminals turned more and more to kidnapping as an alternative source of income. Many families are now haunted by a fear of being caught up in the lawlessness – and local strongmen who promise protection can call the shots. Stephen Gibbs talked to one of them.

San Pedro Garza García is known as the Beverly Hills of Mexico. As I drove through town, I began to see why. In the November sunshine, women in immaculate tracksuits were jogging alongside the main boulevard. They could check their reflections in the shop windows of Louis Vuitton, Cartier and Lamborghini. As the road got steeper, the houses became more impressive still: vast colonnaded mansions, clinging to the mountainside.

I was heading for the house at the top of the hill: the home of Mauricio Fernandez. The recently elected mayor of San Pedro, who reportedly has a fortune of $800 million, last month proposed his own radical solution to Mexico's appalling organised crime problem. His strategy is that the people should not sit back and wait for their government to make this country safer. Citizens, he says, should unite and, if necessary, fight against the criminals.

The suspicion is that the mayor has already heeded his own advice. On the day of his inauguration, he told a crowd of his supporters that he had some good news. A notorious local kidnapper had been murdered, he announced. But he was speaking almost four hours before the man's body was discovered by police in Mexico City. 'I had a tip-off,' is his explanation.

At the gates to his home, a team of leather-jacketed private security guards was expecting me. I was waved on. The entrance to the house itself is through an understated low doorway. The mayor was waiting on the other side.

'It's fifteenth century,' he said. He'd noticed my jaw drop as I gazed at the ceiling twenty metres above us. Vast, finely engraved wooden beams, of the sort you might see in a French cathedral, topped the enormous room in front of us. In the 1970s, Mr Fernandez had bought the entire Renaissance ceiling from the estate of William Randolph Hearst, the American media tycoon, and had it shipped to Mexico.

On the far wall was a large rock, riddled with fossils. 'It looks like Jackson Pollock, but it is thirty million years old,' joked the mayor, his baritone voice echoing around the room. I was handed a shot of tequila. It had been poured into a freshly carved tomato. Along its rim were sprinkled brown crystals. 'That's Hawaiian salt,' my host reassured me. I began to realise that I was in the home of a man whose friends, and enemies, had to admire his attention to detail.

We sat down. Just behind the mayor's head was a dinosaur skull, mounted on a marble plinth. On the table, there were fossils of long-extinct fish. At his feet, there was a tame racoon. But for all the prehistoric and animal life around us, something was missing: family life.

Mayor Fernandez has six children. All are now living in the United States. After a spate of kidnap attempts, first against his daughter, then against two of his grandchildren, he told them to leave Mexico. He says he expects the family to be back within a year or so, after, as he puts it, he has 'sorted out' San Pedro.

His vision is that the town becomes a place where all crime is eradicated. One problem, he admitted, is that utopia has its price, and that among the select people who can afford homes in his neighbourhood, some might have made their money from drugs.

He told me the story of how an affable dad on the local school run turned out to be one of Mexico's most wanted men, a cartel boss with a $2 million bounty on his head. He was arrested last March, along with a stash of machine guns. 'I can't stop them living here, but I can stop them working,' said Mr Fernandez.

This year two mayors have been killed in drug-related violence in Mexico. I asked him if he feared for his life. 'They could kill me today,' he said. 'But perhaps they know it is better not to.' His beautiful assistant stared, a little awkwardly, at the Persian carpet.

Mr Fernandez's popularity has soared in the last month, as Mexicans look for a believable solution to the crime problem which blights their lives. His name is even mentioned as a possible presidential candidate, when Felipe Calderon's term expires in 2012. I asked him whether he has such ambitions. In the great tradition of those who may yet seek high office, he denied it. 'I like being mayor,' he said, as he peered through his Gothic arch at the city below.

Linda Pressly – Guatemala – The Scene of the Crimes – 06/08/11

Just as the problem of coca cultivation ebbed and flowed over borders, so Mexico's cartel wars also spilled over to affect its neighbours. A further hellish dimension to the violence emerged as it became clear that drugs and guns weren't the criminals' only currency. They were also trading another commodity: migrants from Central America trying to move north through Mexico to the United States. People-trafficking was a logical step for the cartels – whether by extorting 'fees' and protection money, forcing female migrants into prostitution, or compelling others into slave labour, the cartels could extract still more profit. And when

*they wished to send a message of intimidation, migrants – undocu-
mented and unprotected – were there to be brutalised.*

There are just so many shoes . . . Flip-flops, a cowboy boot, welling-
tons, a single cheap trainer, and a child's Sunday-best navy leather
sandals. They are strewn across the tropical earth outside two rough
wooden buildings. These would've been home to the workers on
Los Cocos ranch – migrants from the south of Guatemala who
moved north to the region of Peten, near the Mexican border, to
get a job. Now those workers are gone, their bloody, decapitated
bodies removed in refrigerated trucks the day after the massacre.
Twenty-seven were murdered here.

Birdsong carries on the hot, still air. Occasionally I hear voices.
The young soldiers who accompanied us have fanned out across the
property, and are chatting in twos next to empty cattle pens.

Past the wooden homes you reach a clearing. There's a well, and
just ahead a whitewashed brick building with a veranda tiled in
terracotta. I've seen pictures of this scene on the internet, but it's
still deeply shocking. The entire front wall of the building has been
stained with blood in a message to the owner of the ranch written
in giant letters. 'I'm going to find you,' it says, 'and I'm going to leave
you like this.'

It's signed Z 200. In a whisper, our guide tells us the gruesome
details of how the message was written. Extreme, stomach-churning
violence has become a mark of the Zetas – the group believed to be
responsible for what happened here, apparently in revenge for an
unpaid drug debt.

Over the last few years the Mexican cartels have moved into
Guatemala with the aim of controlling territory. Peten is home
to hummingbirds, ocelots and jaguars. It was the birthplace and
heartland of the ancient Maya civilisation, where great cities rose
out of the thick forest. Now it's one of the main transit routes for
cocaine – a huge area of sparsely populated, remote jungle with
little infrastructure, ideal for organised crime to melt into.

The brutality at Los Cocos completely stunned Guatemala – not
easy in a nation where gun, gang and drug violence are everyday

realities. The president arrived within hours of the massacre, and Peten was put under direct military control.

When we visit his headquarters, the colonel in charge of the operation is behind his desk in a room with vicious air conditioning, the speakers on his computer belting out a popular ballad. He offers us cold drinks, and shows me pictures of his wife and granddaughter who're at home in Guatemala City, hundreds of miles away. Then we go outside to do the interview.

The colonel clocks our translator … Matthew is a pony-tailed American who arrived in Guatemala more than a decade ago with the Peace Corps, and then stayed on to set up a local environmental organisation in Peten. He's an invaluable guide for us – he knows the area, is in touch with communities, and is just one of those people you always hope you'll meet on an assignment, but rarely do. But the colonel is suspicious of Matthew. Perhaps he thinks he'll twist his words. 'You're here to translate *not* interpret,' the colonel tells him. And then, realising I'm taller than him – which he thinks might not be a good look on television – he orders one of his minions to bring over a breeze-block, and stands on it. I misinterpret this as quite a good gag, and start to laugh. But the colonel is deadly serious. Then it dawns on him that standing on a breeze-block in front of his men might be an even worse look, and he orders it to be taken away.

The colonel won't be drawn on how many more soldiers have been brought into Peten since the massacre; all he'll say is that the military has the area under control, and that it is working in harmony with other institutions to keep the peace.

The ban on civilians carrying a gun may bring down the number of murders – no bad thing in a region where the murder rate is twice the national average (and Guatemala already has one of the highest national averages in the world). But locals question whether the military does indeed have the area as much 'under control' as the colonel says …

Father Javier, a local parish priest with an easy manner and a cotton shoulder bag, tells me the presence of the soldiers has made no impression on the cartels or the drug dealers. Since the massacre at Los Cocos, the fear is palpable. 'It's always been there,' he says, 'but now it's like the screw's been tightened.'

Katy Watson – Mexico – El Chapo Escapes . . . Again – 18/07/15

Despite the havoc they wreaked, the heads of some Mexican cartels became storied figures – their exploits recounted in song, pulp fiction, film and many a TV mini-series. Tales of their eccentricity, as well as their brutality, were circulated endlessly; everyone knew their aliases, as well as their names. One of the most discussed – though certainly not the most notorious – was the long-running saga of Joaquin Guzman Loera, better known as 'El Chapo', meaning 'Shorty'. As head of the Sinaloa cartel he had ordered countless drug shipments and countless assassinations; he'd also cultivated relationships with corrupt officials and the occasional celebrity. His legend only grew when video emerged showing him calmly leaving his prison cell . . . via a custom-built escape hatch in the shower cubicle. As Katy Watson heard, many Mexicans had grown sceptical of the state's ability – or willingness – to hold him for long.

'It's all a show, isn't it?' said the lady at immigration when I went to renew my visa. We were chatting away about El Chapo as she went through my paperwork. She and her colleague both shook their heads. They don't believe the government's account of his recapture at all. 'Oh, what a cynical lot you are!' I said jokingly – and the immigration officials laughed.

Many more Mexicans have found themselves chuckling this week after details emerged about a series of instant message conversations between the drug lord and Kate Del Castillo, the actress who brokered his interview with Sean Penn. 'I'll look after you with my own eyes,' writes El Chapo, to which she replies, 'I'm very moved, nobody has ever looked after me.'

It was his obsession with the actress that authorities say led to his undoing. His contact with her led them to him. Whether everyone believes it or not, it's certainly another fascinating twist in an extraordinary tale.

Mission accomplished, President Enrique Peña Nieto called it when he tweeted last Friday that the authorities had recaptured Mexico's most wanted man. But very quickly, the jokes started circulating on social media. One, a fake tweet from the president, read: 'We regret to inform you that he's escaped again for the third time.' Another showed the picture of El Chapo when he was captured, sitting on a bed with a dirty white vest on, looking pensive. The caption reads 'Just waiting for my Uber to the airport'.

The problem is, the president, who was already struggling with low approval ratings last year, lost a lot of credibility when El Chapo made his escape. And that was his second, of course – he had disappeared from another high-security prison, supposedly in a laundry cart, back in 2001, although many think the prison officials just let him walk out.

So now El Chapo's recapture *should* be good news, but the government is struggling to be taken seriously. There are many who think El Chapo's just going to do it again. Which is probably why Mexico seems open to the idea of extraditing him. The authorities have already started the process and although it's not guaranteed he'll go – there's a long judicial process to go through – it is a real possibility.

Extradition is a sensitive subject here. Within hours of the news of El Chapo's capture breaking, the local radio stations were full of chat about the pros and cons of putting him on trial in the USA. There are many who feel it would be a big mistake, showing the world that Mexico is incapable of ensuring a dangerous criminal can stay locked up. And of course he's Mexico's most wanted criminal – there has to be some pride in doling out punishment on home soil.

But let's be honest. He's escaped twice now, what's stopping him from doing it a third time? 'The guy's very creative, you've got to give him that,' says Alejandro Hope, a security expert in Mexico City. 'The working hypothesis should definitely be that, given enough time, he will find a way to get out of prison.'

The man is loaded with cash – El Chapo even boasted of it in the interview he gave Sean Penn, saying he controlled most of the marijuana, cocaine and heroin that enters the USA, before going on to tell the actor about his fleets of submarines, boats and planes. And therein lies the problem: he's a billionaire drugs lord being ordered around by prison guards who earn a pittance. The scope for bribery and corruption is great.

But what does the future hold for his Sinaloa cartel? Most experts I've spoken to don't think much will change, assuming El Chapo can still wield some influence from inside.

Mexican government officials have told me that there'll be ten times the number of security cameras there used to be. And there will be an elite team watching him around the clock. He won't even be able to stay in the same cell for more than a few days; he'll be moved constantly to keep him in custody.

Going to the United States to serve out his sentence would make things even more difficult for him. Which is why El Chapo won't want that to happen. As the security analyst put it, 'Right now, for El Chapo, it's a race between extradition and escape.'

Update: after this dispatch was broadcast, Joaquin Guzman Loera was indeed extradited to the USA, where he was put on trial in New York City and sentenced to life in prison.

Will Grant – Mexico – The Convoy – 03/11/18

Fear of the Mexican cartels and Central American gangs was stoked in the United States by news coverage – especially of crimes that their members had committed after entering the USA. President Donald Trump promised not just to 'Make America Great Again', but also to make it 'safe' again. He stressed that tightening immigration enforcement on the US–Mexico border was vital to keep Americans secure. But the pressures driving people north – poverty, insecurity and lack of hope in their home countries – had not abated. Despite Washington's threats to deploy troops to the US–Mexico border to stop them, 2018 saw a number of 'caravans' of migrants from Central America banding together to make their way through Mexico en masse, walking for weeks on end. Will Grant travelled alongside them for part of the way.

If you spend enough time with a group of 7,000 people, it's surprising how often you bump into the same faces. I first met Isaac Perreira, aptly for a migrant, on the road. In the glare of a sun that could melt the tarmac beneath his feet, he was hobbling over a highway bridge in the rural state of Chiapas, a long trail of people in front and behind him.

Isaac has a disability he doesn't know the name of or ever had properly treated. His limbs are stick-thin, with atrophied muscles, and his movements jerky and limited. 'My mother had a fever when she was pregnant with me,' he says matter-of-factly, 'and it got into my bones.' That's all he knows. Now he hopes doctors in the United States might help him improve his mobility.

Slowed by his disability, but by no means one of the stragglers, Isaac still sets an impressive pace. His progress is marked by the rhythmic click-click-click on the asphalt of a cane he's improvised from a broken umbrella. We chat beneath a tree as he takes a few gulps of water while his companion tries in vain to flag down a lift. 'I'm a bit tired,' he says – an obvious understatement – as he pushes up a hat donated by the Red Cross to mop his brow. 'We kept walking even through heatstroke. We're risking everything for a dream,' he acknowledges with brutal frankness. 'But, God willing, we'll get there. We have to finish what we've started.'

Isaac's determination to reach the USA is formidable but not unique. The migrants have spent three weeks on the move since they set out from San Pedro Sula in Honduras amid a wave of optimism. They've crossed Central America, faced down riot police at the border and pushed on into Mexico. And they have suffered. Soaring temperatures, driving rain, sleeping rough, eating little – it's hard enough on the adults. For the hundreds of children, many of them babies, it's even more dangerous.

Isaac unties the brown laces on his trainers and massages his bony, misshapen feet. 'What matters here is the attempt. At the very least, I'll have some stories to tell my brother's kids: a disabled guy doing this!' He grins at the idea.

The next morning, having spent an uncomfortable night camped out in the town of Arriaga, the caravan left before sunrise. A mass of bodies, blankets, rucksacks and prams slipped out of town under the cover of darkness, a sea of empty plastic bottles, rubbish and discarded clothes the only sign they'd been there.

I caught up with them again at the state border between Chiapas and Oaxaca and travelled alongside, walking when they did and catching a lift when local drivers were kind enough to let people on. It was a brief glimpse of their journey, but long enough to appreciate just how punishing the heat is, just how grateful they felt for the occasional helping hand or bottle of water.

At one point, Isaac came past. He seemed to be working harder than the day before, his discomfort clearer. Still, he wasn't about to accept the offer of asylum the Mexican government had made overnight. 'It would be like being in prison,' he said of the stipulation that the migrants had to remain in the southern Mexican states of Chiapas or Oaxaca while their applications were being processed. For that, he figured, he may as well keep going and apply for asylum in the USA instead.

The next town, Tepanatepec, had a river running through it. Within minutes of arriving, hundreds had stripped off and were bathing in the cool water. It didn't look very clean but must have been refreshing after the day's travel.

On the banks, small groups were drying out their clothes or sleeping. Plumes of marijuana smoke wafted past as the caravan's more anti-social element made its presence known. The characterisation by President Trump of the group as riddled with gang members, even Middle Eastern terrorists, is inaccurate. Most are families or mothers with children. Yet it would also be inaccurate to ignore the minority of unsavoury characters among the caravan, a reflection of the culture of street violence many are fleeing.

Isaac hopes they don't ruin the chances for everyone else. Sitting on a rock in his underwear, enjoying the peace, he revealed a tattoo across his heart in swirling letters: 'Naylin', the name of his niece living in New York he's trying to reach. Someone had donated crutches to help him. When I asked him why the USA should let him in, Isaac said he wasn't expecting any miracles. 'It is in God's hands,' he said simply.

Since we'd last spoken, President Trump had announced the deployment of 5,000 troops to the US border. Would that put him off, I asked? 'Not even ten thousand,' he smiled.

Brazil

Justin Rowlatt – Rainforest Farmer – From Hero to Zero – 18/02/12

In the twenty-first century, the fate of Brazil's rainforest became a symbol of the world's relentless hunger for natural resources. The destruction of vivid green jungle, home to myriad species, to make way for endless acres of pasture or soybeans for animal feed seemed to exemplify everything unsustainable about the global food market. During the presidencies of Lula da Silva and Dilma Rouseff, the government – then controlled by the left-of-centre Workers' Party – claimed it had nearly achieved its target of cutting deforestation by 80 per cent. Yet Brazilian policy on the rainforest was once very different: back in the 1980s it was seen simply as open land, fertile for development. Justin Rowlatt reflected on how attitudes have changed.

'There was nothing here when we came here twenty-five years ago, nothing at all,' Gilmar Burnier told me, his eyes twinkling with pride as he led me through Querencia, a small town in northern Brazil. 'Nothing except trees, of course,' he added with a laugh. Gilmar is an easy man to like. He's small and lively with tightly cropped greying hair and a bright, mischievous manner. 'We opened all this up. We cleared the forest and built everything ourselves – roads, houses, everything,' he said.

In truth, Querencia is rather ordinary – a sleepy rural town with grey concrete buildings and wide quiet roads – but to Gilmar it is clearly the most wonderful place on earth. He swept his hands wide, gesturing towards a school playground with swings and a climbing frame made from recycled tractor tyres. 'This was one of the first places we built,' he told me. 'We have three schools now.'

We crossed over to a restaurant on the main street. 'This is owned by an Italian,' Gilmar said. 'He arrived here in 1986 with only a backpack on his back and now just look!' Once again it was nothing fancy – big plate-glass windows with neat rows of Formica tables inside. At the back was a counter piled with salads and barbecued beef.

But Gilmar has good reason to feel proud of Querencia. Its creation is a classic story: hardy pioneers risking everything to make themselves a better life on a dangerous frontier. Yet Gilmar's story, and that of hundreds of thousands of pioneers like him, is rarely ever told. The reason is simple: Querencia is in the Amazon and the town is only here because Gilmar and the other townsfolk have burnt or cut down tens of thousands of acres of primary rainforest.

Back in the early 1980s there was a massive government campaign to get people to come out here. The Amazon was, they were told, a 'land without people for a people without land'. It was a patriotic duty to turn it into productive farmland. That is when Gilmar came up from the far south of Brazil. Like many of the settlers in this region he's a *gaucho*, born into the great ranching and farming tradition established by migrants from Europe. 'We knew how to farm,' he told me, 'we just didn't have any land. So, when we got out here, we did well.'

Just how well he and his fellow *gauchos* had done became clear when we finished up our food and Gilmar took me out to his farm. From a car, you would never know that just a couple of decades ago this whole area had been dense jungle. Now it is vast open soya fields, with only the occasional dark smudge on the horizon where forest still stands.

Once again Gilmar swelled up with pride as he described the huge farms he and the other settlers have carved out of the forest. As far as he's concerned, they have simply done what they were told to do, but his community has faced fierce criticism as attitudes to deforestation have changed. 'Now they treat us like criminals,' he tells me. 'The Environment Agency has come here in helicopters and in trucks with machine guns. They behave as if we are bandits and all the people here were very upset.'

There has been a lot of illegal deforestation around Querencia but even so it is hard not to feel some sympathy for Gilmar. 'We didn't know it was wrong,' he told me defiantly when I asked whether he feels guilty. 'We were just doing what we'd been told was right for us and right for our country.' He claims many farmers are now beginning to recognise the value of the jungle. 'My daughter loves the forest,' he said. 'She wouldn't let me cut any more down even if I wanted to.' He told me he started replanting trees beside the rivers and streams on his land and improving the management of the forest. There is now much less illegal deforestation going on in the area.

So how far might this go, I wanted to know: would his fellow farmers consider reforestation? Gilmar grimaced. 'I know some *gauchos* who would rather go to prison than to plant forest on land they have gone to so much effort to improve,' he said emphatically. Attitudes like this have led to the demonisation of Amazon farmers, but actually most of the world's farmers hold similar views: because most farmland was once forest. Mankind has been destroying the wild places on the planet ever since the first seeds were sown. The difference is that in the Amazon it is still happening right now.

Julia Carneiro – Pregnancy in a Time of Zika – 10/03/16

The Zika virus is not new – it was first identified in the 1950s – and it has been found around the world in tropical climates, with recorded outbreaks across Africa and the Pacific Islands. Like dengue fever, malaria, yellow fever and chikungunya, it spreads among human hosts who've been bitten by infected **Aedes aegypti** *mosquitoes. Its earlier forms seemed to cause no more than a week or so of mild fever, but in late 2015 the Brazilian state of Mato Grosso reported a peak in infections and a terrifying new complication: a greater risk of microcephaly in babies born to mothers who'd had Zika. Microcephaly leaves infants with undersized, underdeveloped brains and varying degrees of life-long disability. Brazilian women were advised not to get pregnant* *during the outbreak, if they could help it, and Julia Carneiro was one of millions contemplating the same dilemma.*

There's a blurred black-and-white shape on the screen and a rhythmic thumping sound suddenly booms from the speaker. Edna Mendonça smiles at the ultrasound screen, her face lit by the monitor in the dark room. Her baby girl is thirty-five weeks old – almost due.

Dr Adriana Melo is gliding the ultrasound probe back and forth over Edna's big belly, silently scrutinising the images. In a soft, barely audible voice, she tells Edna, 'Your baby's head is a bit smaller than it should be,' she says. 'Can you come back to my clinic next week?'

We're at a university hospital in the city of Campina Grande, in the north-east of Brazil, and the room is full of pregnant women waiting to see Dr Adriana. Like Edna, they have all had Zika. But she's the only one that day to leave with a huge question mark hanging over her. Could her little girl have been harmed by the rashes and fever Edna had early on in her pregnancy?

This is the possibility dreaded by pregnant women across the country, and by countless others thinking of having babies. And by me. I had been planning to start trying for a baby myself, later on this year. But suddenly, getting pregnant in Brazil feels like a shot in the dark.

Over the past months, reporting on the Zika story, I've been talking to women about their fears, their plans to delay pregnancy, the impact of having a child with microcephaly. And as I report, I've also been thinking – what on earth am I going to do?

Night has fallen outside when Dr Adriana finds time for me. I ask her: what are you telling women who want to have children? She says those who can wait, should wait. Because no matter how much repellent you use, you never fully protect yourself from mosquitoes. There is always some risk. But some women can't delay for long. 'If they are already over thirty-five years old and have more risk of a chromosome mutation,' she says, 'they can't wait.'

At this point I feel thrust right in the middle of her at-risk group.

It's the day before my thirty-sixth birthday. So she's saying it's risky to try, but it's also risky to wait. It had seemed complicated enough to be planning a baby and a career at the same time.

Next day I go to a different hospital where a special unit for microcephaly has been set up. Everything here is stripped down to the basics, as in most public hospitals in Brazil. The receptionist, sitting behind a metal grille, hears 'microcephaly' and points up the stairs.

Six women are seated at the end of a long dark corridor. Their babies are fast asleep despite the loud chatter. The mothers are joking, gossiping, chatting. They bring their babies here twice a week and are like old friends now. The microcephaly is evident: the infants' foreheads seem to be shortened; their faces seem disproportionately large.

This is a women-only environment – I'm told fathers rarely bring babies here for care.

One of the women, Francileide, says she's been struggling to buy her baby's milk. She now has five mouths to feed, and won't be able to go back to work. Rafael demands all her time. He's three months old and has a frail little body. The doctors fear he's malnourished. I notice his tiny socks with the word 'Hero' written on them. They're so worn out that she has tied a string around his ankles, so the socks won't fall off.

Ianka is the youngest mother here. She just turned eighteen and Sofia is her second child. When she was seven months pregnant, she found out about the microcephaly. Her boyfriend immediately left her. 'He told me he was disgusted,' she says. Baby Sofia is wrapped in a pink blanket with the words 'Princess'. My heart skips a beat when I notice her head. The scalp is a bit loose and wavy; the doctors explain it's because her brain didn't grow enough, so there is too much skin.

There are so many different layers to the tragedy in this room – poverty, teenage pregnancy, women abandoned by their partners. Imagine how many others are having to reassess plans, rethink motherhood, cope with anxious pregnancies. Zika is not only impairing hundreds of children and their families; it's also affecting a generation of women like me who once thought having babies was just a matter of planning and deciding. Now, there's a new factor to consider – and that's truly frightening.

Wyre Davies – Impeach! – 16/04/16

Brazilian politics are often described as complex, corrupt and viciously partisan. The byzantine legal and political drama nicknamed 'Lava Jato', or Operation Carwash, provided plenty of evidence to back that up. An international enquiry originally ordered by a judge, it uncovered a vast network of bribes, kickbacks and money-laundering which implicated several multinational companies and members of governments across Latin America, from Peru to Panama. Brazilian voters might not have been surprised by the allegations, but many were certainly startled by the range and the depth of the financial murk; the country's decade of good results under the Workers' Party seemed seriously tainted. Wyre Davies unpicked some of the knottier strands of the whole affair.

Every time I come to Brasilia, I'm not really sure what to make of the place. Brazil's capital city is unlike anywhere else in this vast, diverse country of more than 200 million people.

Founded in 1960, and deliberately located in the middle of the country, Brasilia was designed by the renowned architects Lucio Costa and Oscar Niemeyer. The city was meticulously planned and set out in the shape of an aeroplane or a bird. Residential and commercial areas are the city's wings while the fuselage or body is where you'll find government buildings and ministries.

At the head of a beast that has long since outgrown its original design and parameters are the presidential palace and, under one roof, the two houses of Congress. It is in the latter building, one of Niemeyer's most striking modernist designs, that 513 congressmen (and they are overwhelmingly men) are currently sitting in judgement on the country's first woman president.

Dilma Rousseff is fighting for her political life in a battle that has become increasingly divisive and hostile. Ostensibly she faces impeachment over charges that she illegally manipulated government accounts to conceal a growing deficit. But the beleaguered president argues that such fiscal tactics have been used by many administrations before and are not serious enough offences to warrant her impeachment and removal from office. This process, say supporters of the ruling Workers' Party, is all about business and political elites trying to force out a democratically elected government. It is, says Dilma Rousseff, a judicial coup d'état.

Ms Rousseff's arch-enemy is Eduardo Cunha. As Speaker of Congress he is meticulously planning the impeachment debate as opposition members try to muster the two-thirds majority vote needed to trigger a formal impeachment trial in the Senate. It's a plot that could have come straight from the dark political television thriller *House of Cards*.

Opinion polls show that Ms Rousseff is a very unpopular leader, her authority undermined by an economy in recession and a corruption scandal that has implicated dozens of politicians and business leaders. The irony, which seems to be lost on many Brazilians, is that many of those politicians debating the charges against their president are themselves accused of much more serious crimes.

Among them is . . . Mr Cunha. He's been accused by federal investigators of hiding millions of dollars in overseas bank accounts – the proceeds, it's alleged, of corruption involving the state-controlled oil giant Petrobras. Eduardo Cunha brushes off the accusations as if they were a mere inconvenience. He's safe in the knowledge that, like all members of Congress, he can only be tried in the country's Supreme Court, where there is such a backlog that cases often run out of time and expire.

What is happening in Brasilia is an unedifying spectacle – leaders of both camps engaged in pork-barrel politics, making promises of plum posts in return for votes. At one stage this week congressmen were pictured openly gambling – laying bets of 100 reais, about £20, on the size of the margin by which the impeachment vote would succeed or fail.

Watching, like most Brazilians, from afar at his family's modest home on a hillside in Rio de Janeiro is Ronaldo Marinho. Ronaldo's father was, and still is, a road sweeper and the family lives in the Pavaozinho favela or shanty town. At twenty-two, Ronaldo is a product of the stunning progress made in Brazil since the year 2000. A country once described as Third World rose to become the world's sixth-largest economy and an estimated forty million citizens were brought out of poverty.

Ronaldo has benefited from a full education and opportunities his parents could only have dreamed of. But the student, who dreams of a career a long way from here, is afraid the country is going backwards. 'The problem with Brazil is that everyone has short memories,' he tells me as he looks down on the upmarket streets of Copacabana and Ipanema.

In recent months these streets have bulged with huge anti-government protests. They are largely, but not exclusively, white and middle-class opponents of the ruling Workers' Party. After a decade of economic growth and unprecedented social progress, Brazil has hit the buffers and people are looking for someone to blame. Many divisions that perhaps never really went away have re-emerged – between north and south, between rich and poor, and along racial lines.

Usually congressmen can't wait to get away from Brasilia – jumping on Friday afternoon flights back to more agreeable cities like Rio, Sao Paulo or Belo Horizonte. But this weekend, all roads lead to the capital. Brasilia may be completely detached from the wider country and the concerns of most Brazilians, but it's here that the fate of Brazil's first female leader and the future of its imperfect democracy will be decided.

Caribbean

Matthew Price – Haiti – Shock and Aftershock – 16/01/10

Dictatorship, demonstrations, deprivation – in Haiti the last half-century has been punctuated by disasters, against a background of crushing poverty, ecological destruction and steady emigration. But the catastrophic earthquake which shook the country on the afternoon of Tuesday, 10 January 2010, was in a different league – not just for its magnitude but for the appalling damage it inflicted on people already struggling to survive. More than three million were affected by the quake and its aftershocks and more than 100,000 died. Landmark buildings – including the presidential palace – were pancaked into ruins. Tent cities sprang up within the cities of Port-au-Prince and Gonaives as survivors tried to salvage what they could of their former lives and keep their families safe. Matthew Price wondered where reconstruction could even begin.

It's a relief to hear laughter amid all the chaos, but Frantz Zepherin can't seem to stop laughing. And painting. And drinking. I bumped into one of Haiti's best-known artists as he arrived at the Galerie Monnin, in Petionville, with his latest work in a plastic bag under his arm. He fished it out: a dark, desperate vision of faces trapped in the rubble, their torment rendered somehow more acute by the fact that a spider has already spun a web across the opening.

But it was an earlier work in his burgeoning earthquake series that had Frantz laughing like a drain. A family of skeletons is marching through the city, dressed in their smartest clothes and treading on footprints that represent, in more ways than one, the souls of the dead. They carry signs, pleading for assistance and thanking the international community. *We need help*, reads one.

Food. Water. Medical. *We need shelters*, reads another. Clothes, condoms and more. Condoms? Frantz let out a long earthy laugh. 'Of course! People are sleeping in shelters, they're making love to strangers. They have to be careful!'

The skeletons, Frantz told me with a giggle, are the living dead, wandering through their city, dazed, like zombies, waiting for help. It's a vivid, accurate description of the state of Port-au-Prince, one month on. It's a city full of zombies. Of people wandering, and waiting. Waiting for someone to come and remove the bodies that still lie, rotting, in the rubble. Waiting for a reliable expert to cast an eye over the ominous cracks in their homes, to decide what stands and what must go. Sitting, waiting, drifting. With dull, exhausted expressions. The living dead.

To be sure, there's plenty of enterprise too: someone selling Haiti's ubiquitous paintings from the tent in a park where he now lives. A couple of mechanics, slowly scavenging anything remotely useful from the wreckage of a crushed, mangled car. A team of workers, armed only with hammers and perched precariously on top of a badly damaged building, laboriously chipping away the concrete in an effort to preserve the ground floor.

But these are pockets of industry in a city still crushed by the weight of falling concrete and appalling suffering. Haiti, of course, is ill-placed to dig itself out of its physical and psychological mess. When a larger earthquake hit Kobe in Japan, fifteen years ago, the Japanese government invested $58 billion over four years to repair the city. The Haitian government can muster a mere £1 billion a year at best. Unless the international community shoulders the financial burden, then it's hard to see how Port-au-Prince can avoid the fate of the Nicaraguan capital Managua, where large swathes of the city centre remain desolate and overgrown, thirty-eight years after its own devastating quake.

With the rainy season due in a couple of months and the very real danger of hurricanes later in the year, Port-au-Prince is desperately in need. In Rue Valiant – Valiant Street – I found people still traumatised and passive, living in a long line of white tents down one side of the street. They made occasional nervous forays into their damaged homes but were much too scared to stay inside for long.

Veteran community volunteer Emeline Desert – still sleeping in her jeep – told me the problem was one of organisation. Haitians are strong and creative, she said, but they're not educated to participate. Nor did they listen to those who warned that building flimsy homes in an earthquake zone was inviting disaster.

In the absence of anything more organised, Haitians are starting to turn their makeshift shelters into semi-permanent dwellings. Settling down where they can, thinking only about making it through another day. Reluctant, according to an Oxfam survey, to move to camps away from Port-au-Prince. Before the quake, experts pondered ways of trying to encourage people to move out of a dangerously overcrowded city. When it struck, some wondered if nature hadn't just given a cruel push. But for now, the living dead are mostly staying put. And waiting.

Nick Davis – Jamaica – Simply Murder – 06/12/12

Journalists in Jamaica often express resignation – or resentment – at the way their island is portrayed in international media. And it's true that coverage does focus heavily on the country's crime problem, with occasional bursts of enthusiasm for its achievements in music or sport. Striking the right balance between hard-hitting investigation and sensationalism can be tough – especially when cops, as well as robbers, are often keen to emphasise their willingness to use violence. But Nick Davis was still shaken by what happened to a friend.

Until I open my mouth, people here take me as being a son of the soil. My parents are Jamaican and, despite being born in the UK, I still have a sense of patriotism that sometimes leaves me walking an editorial tightrope. Like all Jamaicans, I hate the negative publicity that's too often associated with this island.

The biggest issue I face is reporting crime. In the past, the BBC as well as other international broadcasters have been accused of parachuting in reporters from abroad to cover stories here with little context, giving a distorted view of this country and its problems. Looking at some foreign coverage of Jamaica, you might conclude that there's nothing *but* crime going on here.

The reality of living here in the long term gives a truer sense of what it's like on the ground. And the truth – as unpalatable as it may be for me and others – is that crime in Jamaica is still a problem that can't be ignored, even though things have got better recently.

Last week saw the funeral of thirty-year-old Tandy Lewis. Her burnt body was found in bushes close to the airport in Port Royal; she was four months pregnant. A few days before her remains were found, I'd been talking to one of her friends, who mentioned her disappearance. She'd not seen Tandy for weeks. Although we tried to remain positive, there was a lingering sense of dread as we spoke.

The police in Jamaica have made great gains since a state of emergency in 2010 that saw Christopher 'Dudus' Coke, a feared drug and gun smuggler, extradited to the USA. The following year, 2011, saw the lowest murder rate since 2003. And the official crime statistics show that wasn't just a blip. Between January and September 2012, there's only been a slight increase in the murder rate, up 1.1 per cent year on year, with 31 deaths per 100,000 people. It's an overall downward trend, and so far investigators have cleared up over 300 cases.

Unfortunately, my friend Roberts won't be one of them.

It wasn't till after his death that I came to know his first name was Anthony. He was a taxi driver who I rode with most days. The fifty-two-year-old didn't look his age; he had shiny dark skin, so dark the other drivers teased him for it, calling him 'Ackee', because his complexion was as black as the seed of the ackee fruit. He was an honest man, a hard worker who was also fair-minded. He'd brave the traffic to come and pick me up in the rush hour, or find me in parts of the inner city where others might have left me stranded. He'd come for me in the early hours, or late at night. He spoke about his son constantly, proud, because his boy was a good student. He worked to provide for his family, and always had a smile; he saw the positives in everything.

He was shot dead by a man who robbed his taxi just before Christmas. The 'customer' rang the taxi base and waited to ambush him for his takings. Ackee hadn't even been assigned to the call, I was told, but he didn't want to leave a client waiting on the street, so picked the man up because he was close.

He was the first driver at the company to be killed in over a decade, his friends say. The gunman who murdered him was himself killed in a shootout with the police only weeks later. This is no comfort to Ackee's wife, left without a husband, his son without a father.

But everyone comes to deal with murder like this here. I caught up again with Tandy's friend after her death was confirmed. I gave my condolences, but she simply said she was coping and that she's used to it by now. 'Everybody moves on so quick – they die and we move on,' she told me. That seemed a cold statement from someone I know so well. But then I realised that that's exactly what I've done myself, with Roberts.

Part VII: THE WHOLE PLANET

The Whole Planet

Natasha Breed – Kenya – The Lion's Roar Falls Silent– 07/07/12

While some places struggle to contain or eliminate invasive species, others are fighting to keep their emblematic plants and animals alive. A century ago there were around 200,000 lions roaming the African continent; their populations have been whittled down not only by hunters but, more drastically, by habitat loss, poisoning and lack of prey. Natasha Breed considered how competition for resources – space, water, food – with ever-growing human populations has put people and predators at each other's throats.

The sign at the entrance to my neighbourhood read: 'To All Our Dear Residents – Lion seen, Mukoma Road'. The area borders the Nairobi National Park, and we're used to hearing lions calling at night (sometimes leopard and hyena too), but still, the idea one might encounter a lion while opening the gate was rather thrilling. It felt a bit like forty years ago, when a cricket match at the local school had to be abandoned while the delighted children watched from the safety of the classrooms as lions strolled across the pitch.

When I returned six weeks later, there was a new sign: 'Be Aware – Lioness along Mukoma Road' – along with an emergency number to call. Evidently somebody called the number, as action was taken. One lioness was shot, and another, together with four cubs, was taken to the Nairobi Animal Orphanage.

And then last Thursday's newspaper headlines screamed in unison: 'OUTRAGE AS MORANS KILL SIX LIONS IN NIGHT OF TERROR'; 'DEATH OF LIONS A BIG LOSS'; 'KING OF JUNGLE DOWN AND OUT'; 'FURIOUS

VILLAGERS SLAY SIX LIONS'. 'HASIRA HASARA', yelled one of the Kiswahili papers, meaning 'The Damage of Wrath'.

While each editorial differed, the photographs all told the same story. Grisly images of the limp bodies of lionesses being hefted into a Park Service vehicle by uniformed rangers while a spear-wielding crowd looked on; lurid pictures of punctured, blood-stained animals being exhibited, heads lifted by the ears, eyes staring unseeingly, tongues lolling from gaping mouths. Two of the lions were cubs.

In the early morning, following an attack on a flock of sheep and goats, furious villagers, armed with traditional weapons and using vehicles to herd the marauding lions together, had made a distress call to Kenya Wildlife Service and held the big cats at bay until rangers arrived. The rangers tried to placate the angry men, who demanded a veterinary officer be brought to sedate the animals and take them away. After an hour, the villagers' patience finally ran out, and six lions were attacked and killed. Two others escaped.

Historically, after rains, lions leave the park following herbivores, which disperse into outlying areas. Not many years ago, the migration of zebra and wildebeest between Nairobi and Amboseli National parks was second only to the great Mara-Serengeti migration. But where there used to be a 'wildlife corridor' of land where both prey and predators could spread out, now human habitation blocks the animals' way. Inevitably, predators encounter livestock, and subject heavy losses on pastoralists and farmers alike. To a stockman, the loss of any livestock is drastic, the danger of coming into contact with a predator while trying to protect one's herds considerable.

Paula Kahumbu, chairperson of the Friends of Nairobi National Park, makes regular visits to the adjoining community. She tells me pastoralists have reported frequent attacks on their livestock in the past six months and demanded compensation for loss of their animals. While there is a government policy in place to pay compensation for loss of life or injury to humans caused by wildlife, none exists to replace lost livestock.

This incident serves to highlight the challenges faced by lions and other wildlife in Africa today. Combined with the burgeoning human population, land use policies are changing. Space is precious.

There are now thought to be only 20,000 lions on the continent. Lions survive by moving. As zebras, wildebeests and antelopes follow the rain for fresh pasture, predators follow the herbivores. But in Kenya, with human habitation spreading out across the country, the animals' habitats are becoming no more than islands in a sea of humanity.

Is there really room for lions in developing Africa? Naturally the welfare of humans takes precedence over that of wild animals. Yet, in Kenya, tourism is one of the main sources of revenue, contributing significantly to the economy. And there's surely no beast more evocative of Africa than the majestic lion.

According to Maasai legend, when a lion roars in the darkness across the savannah, he is saying, 'Whose land is this? Mine, mine, mine . . .' But it isn't – not any more.

Rupert Wingfield-Hayes – Indonesia – Sumatra Burning – 29/06/13

Every dry season in South-East Asia, there's a noticeable worsening in air quality as the smoke from uncountable acres of burning rainforest fills the skies. The 'haze crisis' of June–October 2015, which caused premature deaths across the region and made face masks a common sight in major cities, was a dramatic manifestation of the problem. Malaysia, Singapore and Indonesia were all blanketed in yellowing, toxic air for weeks. The Indonesian president was forced to make an official apology for the smoke, and the fires which had caused it. But Rupert Wingfield-Hayes had looked into the root causes of this recurrent fug years earlier.

Someone once told me 'there is no point in worrying about something that you cannot change'. It's good advice, and it means I should probably stop worrying about the destruction of Indonesia's rainforests. But it's hard, especially after what I've just seen.

I first went to Sumatra nearly twenty-five years ago. I visited an orang-utan reserve outside the city of Medan. From inside a huge cage the great orange apes stared out at me with their big intelligent eyes. I remember feeling pain and anger at the loss of their forest home. I feel it now.

In 1989 I cycled across the island and much of it was still covered in a thick layer of rainforest. The outside world woke up – briefly – to its destruction in 1997, when smoke from huge fires cast a pall of smog across much of South-East Asia. Sixteen years have passed … and we are talking about it again. In the meantime most of the forest I saw has gone, and with it much of Sumatra's extraordinary wildlife.

The speed of the destruction has been astonishing. In the mid 1980s there were still around 5,000 Sumatran elephants. Half that number survive. Thirty years ago there were over 1,000 Sumatran tigers; now there are less than 400. Half the orang-utans have also been wiped out. All three species are now critically endangered and heading for extinction.

The timber barons once drove the destruction, but now there is a new gold rush – palm oil. The oil palm is an amazing tree. It is easy to grow, matures in less than five years and is extraordinarily productive. In the wet season it will bear fruit every two weeks, in bunches of large yellow berries, each bunch weighing up to fifty kilos. No wonder it is so popular.

Sumatra covers an area more than twice the size of Great Britain. The whole thing is being turned into an enormous mono-culture. As I drove across the island this week, for hour after hour, all I saw on either side were massive plantations, neat rows of hundreds of thousands of oil palms.

In my hotel in the city of Pekanbaru, I switched on the TV and flicked to the international news channel. Up flashed beautiful images of lush forests, crystal-clear waters and happy young faces; over the pictures came a silky voice. 'Indonesia,' it said, 'committed to sustainable development.' I turned to the window and looked outside. The city was all but invisible behind the thick layer of choking smog. I didn't dare open the window.

Summer in Sumatra is fire season. Once the big trees have been cut for timber, fire is used to clear the land for cultivation. It's completely illegal, but nothing is done to stop it. Locals say the fires are set by agents hired by big palm oil companies from Malaysia and Singapore. The companies deny it.

As we drove through the smog, my local driver tore into Indonesia's wealthy neighbours. On the news, the Malaysian and Singaporean governments had both been complaining about the pollution. 'They come here and they destroy our country,' he said. 'Now they get to taste the results for themselves.'

And there's a final irony. One of the biggest drivers of the palm oil industry is the demand for bio-fuels. Palm oil makes excellent bio-diesel, because it burns cleanly and has a very high energy content. But according to a leaked European Union report, making bio-diesel from palm oil is even worse than using crude oil. On the 'nasty index' of things you definitely don't want to use, it is second only to oil from tar sands. So next time you fill up your car with bio-diesel, do remember: just because it has 'bio' on the pump doesn't mean it's good for the planet.

Sue Branford – Brazil – Power Struggles in the Amazon – 11/12/13

It's tempting to suppose that energy from renewable sources – hydro-power, wind farming, solar panels – is a no-lose proposition. Yet even if these technologies do offer us a way out of a fossil-fuelled dead end, while sidestepping public fears about the safety of nuclear energy, they still come with social and environmental costs of their own. In Brazil, Sue Branford found that plans for a large number of hydroelectric power stations throughout the Amazon basin were causing considerable dissent – especially along the Tapajós River, home to the Munduruku Indians, one of Brazil's largest indigenous communities.

The first time I saw the two little Indian girls I was in their village, called Boca das Tropas, on the Tapajós River, one of the main tributaries of the Amazon. I was sitting on the riverbank under a large tree, sheltering from the sun. A group of children, about fifteen of them, came running down. The older ones dived in from the bank, chattering and shrieking with laughter, while the younger ones walked down to the beach. The older of the two girls – she must have been about six – was carrying the younger one, aged about two. She carefully put her down and held her hand as she entered the water. There wasn't an adult in sight. For the next half-hour the children played happily. I didn't hear a cross word.

The peacefulness of the scene was impressive. Ironic, I thought, as I'd come to talk about conflict – more precisely, about the Indians' opposition to the hydroelectric power stations planned for their river. All the 12,000 Munduruku Indians living by the river will be affected. Earlier this year angry Indians kidnapped three researchers carrying out environmental-impact studies. Now the government has sent in armed guards. Tension increased recently, when one of the Indians' most sacred sites was destroyed by a construction company. It was where the Indians buried their dead and where, they said, there was a gateway to another world.

I walked back into the village. Carrying baskets on their backs, women were cleaning up around their palm-thatched huts. After a while, two of the women sat down with me. They were *guerreiras*, warriors, they said. 'They think they're going to intimidate us but they never will. We are fighting for our people, our children, our nature. We have to save all this.'

In the evening, as I took the boat back to the town of Jacareacanga, I thought how hard it is to reconcile the interests of a powerful government, keen to develop a country's natural resources, with those of an indigenous population. The Munduruku Indians have been in contact with the outside world for over 200 years. But their problems began fairly recently, when geologists discovered vast reserves of gold and other minerals deep below the surface of their land. The Brazilian government and big mining companies want to exploit this wealth, but to do it they need energy.

The government's decision to opt for hydropower is controversial. Even by the high standards of the Amazon basin, the Tapajós valley has an extraordinary biological wealth. It has over 600 species of bird, for instance. One of these, a tiny hummingbird, known as the Tapajós Hermit, was only discovered in 2009. It has 161 species of mammal, compared with 222 for the whole of Europe. Much of this wildlife will clearly suffer with the construction of the dams. Some engineers are arguing for less damaging forms of energy, like solar power, but the government says the technology is untested. While the argument goes on, the government is pushing ahead with the construction.

A fortnight later I saw the same two indigenous girls, this time in the town of Jacareacanga. With designs in black paint all over their bodies, they had come with their parents to protest against the holding of a public meeting about the dams. The Indians insist that these meetings should not be held until they have been properly consulted in their villages. Indeed, the law requires this, and some lawyers from Brazil's public ministry are fighting hard on the Indians' behalf to stop the whole process. But, in the meantime, the government is pushing ahead with the public meetings.

There was considerable tension. The government officials flew into the nearby military base and arrived with a strong military escort. The Indians tried, somewhat ineffectually, to stop them going into the hall.

In the event, the meeting was a sorry affair. It began with the singing of Brazil's national anthem. Ten white men on the stage sang lustily, with the support of the first three rows of businessmen, civil servants and so on. Behind them were Munduruku Indians and people of indigenous origin, standing with their mouths closed in a silent protest. The meeting was firmly chaired – all questions had to be written, with no spontaneous contributions from the floor, despite the fact that many attending were illiterate. I felt as if I was witnessing the takeover of the town by an occupying power.

By the time I left the meeting, the two Indian girls had disappeared. They'd probably gone back to their village with their parents, I thought. Tomorrow they'll be playing happily by their river again. But, as the economic frontier reaches their land, their future looks increasingly bleak.

Claire Marshall – USA (Alaska) – A Cellar Full of Whale – 22/07/17

The impact of climate change on Arctic wildlife and culture is all too obvious. Across the region, the duration and pattern of winter freezes have been changing – making long-preserved survival strategies less reliable. Polar bear numbers have fallen in many areas as fast-melting ice makes hunting harder for animals. On land, Claire Marshall observed how indigenous communities with their own ways of lasting out the winters were having to change their plans.

It's awkward. I love whales and an Inupiat hunter is standing in front of me offering me a piece of one to eat. It's a rectangle of soft white, edged with dark black, about the size of my thumb. 'It's a small piece,' he says, 'like I've cut it for my child – half fat, half skin. Just let it melt on your tongue for a bit, like chocolate.' His eyes close a little – rather dreamily.

This piece has been sliced off a boulder-sized hunk just hauled up from an ice cellar. It's a cave in this family's back yard, which they've spent years digging out from the permafrost. It's basically a freezer that you don't need any electricity to run. This town is so remote that it's entirely cut off by road from the rest of America, and things, including electricity, are expensive.

I look down into the shaft. The only light comes from a single yellow bulb. There are deep red masses glistening with frost, squatting alongside oily grey-white slabs. Here and there, the rubbery black of whaleskin. There's a man in rubber boots, standing on four tons of frozen meat and hacking at it with a pick. 'Watch out for the liver,' he says. 'Don't stand on that.'

After the locals caught a nine-metre bowhead whale in spring, it was cut into pieces and stored down here. Now they're bringing pieces up for a community feast. This isn't just a symbolic tradition, but it's not commercial either – the meat can't be sold. This is subsistence. There is no way of trucking anything, including food, into this town. During the long Arctic winter when the sun doesn't rise, it's important to have your larder stocked.

The pick makes a dull thud as it bites into the frozen flesh. The walls are thickly coated in a sparkling layer of ice. But the problem is, some of these cellars aren't working any more . . . they are melting. The Arctic is warming twice as fast as anywhere else on the plant. Naomi Ahsoak, a young whale hunter, shows me her family's ice cellar. The shaft is surrounded by water. She says it's melting fast; when she was a kid none of this water was here. A friend climbs down and stamps around in the slush.

Climate change isn't just altering how the Inupiat store their meat; it's also changing how they hunt.

I run my hands over the hull of one of their boats. It's thick and smooth and creamy coloured, with lines of slightly sunken stitching. It's made from the cured skins of bearded seals; the basic design hasn't changed for thousands of years. Naomi tells me it's an honour to help make one – she once spent an excruciating forty hours straight stitching one together with a group of other women. It looks so delicate. But six men – and women – go out in these things to hunt the second-largest mammal in the world, the bowhead whale.

During the spring and autumn, when whales migrate, the hunters head out onto the ice and spend days carving a safe way for the boat to be hauled out to the open water. If they harpoon a whale, they have to rope it and pull it up onto the ice – but each year this gets more dangerous because the ice is melting earlier. Naomi's father, one of the most experienced whaling captains in Utchuapick, tells me that all the old signs they once used to read the ice can't be trusted any more.

Not far from Naomi's house science is being done that backs all this up with hard data. At the end of a snow-covered track at the northernmost tip of the town is an observatory. The air is so pure here that it's perfect to monitor the atmosphere. Bryan Thomas runs the lab. What he says is disturbing. He likens climate change to a runaway speeding car that we have set in motion and that we now can't control. Quietly he tells me, 'All we can do is adapt.'

So, as I look at the piece of whalemeat in my hand I try to think of the bigger picture. I do manage to eat it. Unfortunately, though, I don't manage to smile.

Peter Oborne – Bangladesh – The Salted Earth – 09/06/18

While the population of Bangladesh is rising – 163 million people and counting, all living in densely populated, low-lying terrain – the amount of arable farmland is shrinking, thanks to rising sea levels. For decades the country has struggled with frequent flooding, cyclones and typhoons – as well as the threat of famine after crops are lost. According to the Intergovernmental Panel on Climate Change, fifteen to twenty million Bangladeshis could find that their homes are underwater by 2050. Peter Oborne saw – and tasted – what that means on the ground.

Only about eighty Bengali tigers still prowl the Sundarban mangrove forest in coastal Bangladesh, down from 400 ten years ago. Tides are higher, cyclones more frequent, human intrusion destructive. The Sundari trees, which gave the forest its name, are dying or already dead.

I took a boat through the Sundarban: you could see the gaps in the trees. Fishermen told me they have to lift their homes with plinths in order to cope with higher tides. We disembarked at Gabura, around twenty-five miles inland from the Bay of Bengal. Few foreigners come here, and people soon gathered. Fifty-two-year-old Mohammed Arazad told me, 'The weather has changed. The trees are not growing. Something is burning them from the top.

'When I was a young man,' he went on, 'we used to have six seasons. Now there are just three – summer, winter and the monsoon. The tides are more than half a metre higher than when I was young.'

People store water in open pits, but these become unbearably saline within days as it evaporates. They have to drink it anyway and the health consequences are horrible. Villagers suffer from high blood pressure from the salt, and heart disease. Eye problems, skin diseases, diarrhoea and cholera are chronic.

Twenty years ago, the nearest source of fresh water to Gabura was only a hundred yards away. Now people have to travel more than a mile to get to it. They have to wait for hours in queues. There are regular fights about access to what villagers longingly call 'sweet water'.

They say climate change has a taste in Bangladesh. It tastes of salt. One fisherman, scarred from a tiger attack (there are scores of 'tiger widows' in this part of Bangladesh), told me that the rivers, always brackish, have become more salty. The freshwater fish have vanished. There are no more carp or murrel – until recently a local delicacy.

All the villages around here used to grow rice and vegetables and breed cattle. 'Nowadays we can't because of the salt,' said Ana Rani, from Magura Kuni village. 'Even the dates are salty. Even the coconut water is salty.' Some try farming shrimp. But they complain it's getting too salty even for that. And shrimp farming increases the saltiness in neighbouring areas, damaging farming yet further.

Ana Rani told me the mud with which they build their houses is weaker too, because of the salt. The walls crumble and crack, needing daily attention and extra wooden supports. There used to be two crops a year. Now there's only one and for the rest of the year the fields are barren.

Sixty-five-year-old Nasir Uddin Morol from Datne Khali village told me, 'I was born in this place. At low tide we could cross the river by foot. That's not possible nowadays. Forty years ago, when the bananas ripened there were too many of them. You could think of selling them. Now the banana trees are dead. All these changes have happened in my lifetime.'

Some give up the struggle. The men leave first. The collapse of farming means they have to find work elsewhere. Many go to Dhaka, where they live in the slums and work in garment factories or become rickshaw drivers. They are among the world's first climate change refugees. Millions more will follow in coming years.

The villagers understand that they may be forced to leave the places where their ancestors lived. Selina, married with two children, told me, 'We don't know how to live anywhere else. We understand this environment. We don't know how to survive in Dhaka. We must fight to stay here.' But scientists say that with rising sea levels, they have no hope.

These Bengali villagers are tough, hardy and practical. They won't be defeated easily. Selina said she had been born relatively wealthy, but she was married off aged twelve into a very poor family. 'I would steal coconuts at night, I was so hungry. By day I would beg for rice in the streets.' Fifteen years ago she took out a small loan from a local charity. Today she owns about three acres of land and rents out eight shops in the local market. Her husband's back from work elsewhere and they have two children.

Selina, who is starting to dabble in local politics, knows the score. She looked at me: 'You are using air conditioners, cars, household appliances – when we don't even have electricity. *You* are producing the climate change. But it's us who are facing the impact.'

Vin Ray – USA – The Drone Warriors – 07/01/17

The virtual world is not armed with weapons of mass distraction alone. As drone technology became ever lighter and smarter, it extended the reach of missiles and spy cameras far beyond previous limits. It's become easier than ever before for rich countries to wage war from a safe distance. But what about the military men and women using these virtual tools? Vin Ray was given rare access to the only US Air Force base devoted entirely to flying drones, where he discovered the strange double life of their operators.

If you're a drone pilot, there's a strong possibility you live in Las Vegas. And your commute to work is against the traffic. We were told to drive north-west out of the city on Interstate-95. The road stretches out through the barren, inhospitable scrub of the Nevada desert. Pay attention, we were told, because the signpost is small. In fact, it's very small. But we eventually arrived at our destination:

Creech Air Force Base – a small, flat, city in the desert. And the only US airbase devoted to flying drones.

Inside the base, comparisons with science fiction are hard to avoid. A drone looks like a conflation of a giant insect and a light aircraft. It's unmanned. Standing by a runway, we watch a drone land and pass right in front of us. The camera underneath its chin swivels quickly sideways and looks right at us: someone, somewhere on the base, is watching us.

I'm escorted through a nondescript door in the side of what looks like a beige metal shipping container. It's cramped inside. At the far end there's a pilot, seated on the left, who flies the drone and fires the missiles. The sensor operator sits on the right – they operate the camera and fix the laser on the target for the missile to hit. They're focused on a bank of screens, switches and buttons. This is today's kind of cockpit. But it doesn't feel like a battleground.

For a start there's a sensory deficiency. From my experience on the ground, you can taste war – you can smell it and you can certainly hear it. In here there's a just a mute video feed. But that's not the only difference.

Traditionally, soldiers in a war zone are based together. They have each other's camaraderie, and they're separated from their families. It's not the same if you're commuting to work every day.

The drive itself is simple. But the psychological journey is altogether different. Imagine. Between six in the evening and six in the morning you might collect your kids from football practice, pick up some milk on the way home and barbecue the dinner. But between 6 a.m. and 6 p.m. you have a licence to kill.

This commute is familiar to Lieutenant-Colonel Matt Martin. He's a hugely experienced former drone pilot. He exudes a quiet strength and a ready charm. But he talks about his schizophrenic existence, his inability to have a normal life and the strain it took on his family. 'It's a surreal enterprise,' he says. 'You only have the drive to work and then you're flying. So personally, I would take that drive to switch gears. I would step into my cockpit and be totally immersed in flying the drone. Then a few hours later I would step out and be back in Las Vegas, in a totally different time zone, different time of day.'

The base commander Colonel Case Cunningham told me, 'When they walk through the gate, they're in a war. Although physically they are at home, mentally they're at war. So in effect we're asking them to redeploy every single day, to go back home and be parents and be loved ones – and then come back to war again.' Such are the new dilemmas of the modern battlefield.

These drone pilots can sit in Nevada and watch a potential target 8,000 miles away for months on end, gathering what they call 'patterns of life' and building what's been called a 'remote intimacy' with their prey – with the knowledge that, one day, they may kill them.

Then there's the aftermath. A conventional fighter pilot will fire missiles and then head back to base. But drone pilots are required to circle for some hours afterwards, to assess the damage. The picture they're looking at is extraordinarily clear – and the damage is often in the form of dismembered body parts. Small wonder that Creech now employs a psychologist for drone pilots suffering stress. Drones are globalising the battlefield, blurring the boundaries between home and war.

As we get ready to leave the base, the moon rises over the mountains and darkness falls quickly. There's a long traffic jam as some of the three and a half thousand air staff wait at the gates to leave the base – a long snake of red tail lights heading back to Vegas and the warmth of their families. And when they get home? Well, friction can stem from one simple question: 'How was your day?'

Jemima Kelly – Russia – Kaliningrad Bitcoin – 20/10/18

It's not only war which has become increasingly virtual – so has money. Technological innovation allowed for the invention of crypto or virtual currency, which unlike 'real' money held in cash or accounts, is not controlled by governments or a bank. Although different cryptocurrencies have fluctuated in value, they've made huge inroads. Bitcoin, the world's leading variety, has relied heavily on being largely unregulated and anonymous – which makes it highly attractive to criminals. The global value of bitcoin relies on a kind of shared, digital, public ledger to record all transactions, stored on a huge network of computers around the world. Users who allow their computers to be part of that network are occasionally rewarded with bitcoin to keep. As a result, huge bit-coin 'mines' have been set up to cash in wherever electricity is reliable and cheap. Jemima Kelly visited one in a former Soviet enclave on the Baltic Sea.

I'm in a black BMW, hurtling down the motorway towards the air-port in Kaliningrad. I notice – nervously – that the speedometer is showing over 200km an hour – so I try to focus instead on the flat, untilled wasteland stretching all around us. I only met the driver, Andrei Mironovich, a few hours ago, but I was drawn in by his story, and now I'm about to miss my flight.

Amber, nicknamed 'Baltic gold' – *yantar* in Russian – has been gathered in what's now Kaliningrad for several thousand years. But these days, that's not the only kind of gold that's mined in this odd corner of Russia. Some shrewd tech-savvy types have recently moved on to mining 'digital gold' – a term used to describe bitcoin and other cryptocurrencies.

Andrei Mironovich, who's sporting a Louis Vuitton-emblazoned T-shirt and an Apple Watch, is in on this twenty-first-century gold rush. He used to sell amber, but bitcoin mining, he tells me, is more profitable.

When I meet him, it's not in the circumstances I'm used to in the shadowy world of crypto, where the lines between what's lawful and what's not are rather blurred. We're formally introduced by officials in the Kaliningrad Region Development Corporation, a state-owned entity based in a leafy part of town.

The corporation was set up to promote business in this Russian exclave which, surrounded as it is by hostile neighbours and hun-dreds of kilometres away from the rest of Russia, has to be a bit creative with what it promotes.

Inside the corporation's building, I'm with the region's IT and communications minister, who doesn't seem concerned that bitcoin has yet to be regulated at the state level. 'Whatever is not illegal is legal,' he tells me, with a wry smile. It strikes me that Kaliningrad is very much open for business, and it doesn't seem to mind what form that business comes in.

I assume the man sitting opposite me with the slicked-back hair is a junior member of staff, but suddenly he interrupts the minister to explain why this region, with its cheap electricity, cool air and plentiful space, is such a perfect spot to mine bitcoin.

This is Andrei. I'd been hoping to meet a bitcoin miner in Kaliningrad, but hadn't found any. Now, just as I'm supposed to be heading to the airport, I've been brought one by a government minister! Andrei generously offers to whizz me to his bitcoin mine and then on to my flight, so I decide to chance it.

We speed through the centre of the city, with its strange juxtapositions. Newly rebuilt Hanseatic-style architecture, harking back to the days when this was the German city of Konigsberg, sits alongside dreary grey Khrushchev-era tower blocks. As we pull into a run-down courtyard, Andrei's shiny new car contrasts sharply with the old bangers parked alongside it. A delicious, pungent aroma wafts through the air, reminding me that I haven't eaten since breakfast.

We walk into a dilapidated old Soviet factory, to be met by a loud whirring sound. This place is nothing like the high-tech, pristine bitcoin mines I've visited before. A hundred computers hide behind tangles of wires and huge fans, sitting higgledy-piggledy on hammered-together wooden pallets, their flashing diodes throwing light onto the peeling walls.

Andrei says this mine and another one he has across town earn him and his brother about two bitcoins a month – around US$12,500 at current values. That's not bad in a region where the average monthly salary is 31,000 roubles or $450. And he's recently ventured into waste disposal too. He displays all the stereotypical signs of wealth but, like Kaliningrad itself, doesn't seem too picky about where it comes from.

As we walk back through the courtyard, a man in a grubby apron comes over with a white paper bag. His face and accent give him away: he's not from Kaliningrad – or even Moscow – but from the

Central Asian republic of Uzbekistan. 'This is for you: a *samsa!*' he says, handing me a traditional triangular Uzbek pie, from his restaurant at the edge of the courtyard. As we get into the car, Andrei tells me, 'Me and my brother, we often say that we have here a kind of hash of modern Russia – there's an old industrial Soviet factory, there's bitcoin mining, and there are poor Uzbek migrants cooking food.'

When we arrive at the airport, I have just enough time to bite into the pie before going through security.

Sally Hayden – Libya – The SIM Card Is Our Life – 11/10/2018

Audiences brought up on the papers, radio and television are often sceptical about how much social media platforms have really done for news reporting. They might offer the sort of first-hand testimony and images which professional journalists might try and fail to find for weeks. But if a reporter's not there in person, can a story still be honestly told?

As militias occupied much of Libya after Gaddafi's overthrow, there were accounts of terrible abuses committed against would-be migrants. Many were locked up in filthy detention centres after being picked up on the streets or intercepted by the Libyan coastguard as they tried to cross the Mediterranean to Europe. Amid a patchwork of militia fiefs, getting into to these centres is hard for journalists, but as Sally Hayden learned, that doesn't mean those held inside have given up on trying to get their stories out.

I was sitting at home on a Sunday evening when the first Facebook message arrived. 'Hi sister Sally, we need your help,' it read. 'We are under bad condition in Libya prison. If you have time I will tell you all the story.'

The sender explained that his brother knew my reporting on Sudan, and had found my contact details online. Along with hundreds of other people, he was trapped in Tripoli's Ain Zara migrant detention centre, where serious fighting had broken out around them.

'We see bullets passing over us and heavy weapons in the street,' he typed, before sending me some photos which he said were recent. Vehicles with anti-aircraft guns visible outside the centre's gates. And an image of himself: an emaciated-looking man with three young children.

Along with tens of thousands of others, the person messaging me had been returned to Libya by the EU-backed Libyan coast-guard, which has been intercepting migrants and refugees trying to cross the Mediterranean since last year. Like many of them, he was sent back to indefinite detention, in conditions condemned by human rights groups.

Now the detention centre itself was in a war zone. The fighting that broke out in August has been the worst in Tripoli for years, forcing the UN Refugee Agency to concede Libya isn't a safe country to return refugees to, while Médecins Sans Frontières is calling for the detainees to be taken somewhere safer.

As I asked for more and more detail, the Eritrean I was speaking to told me how – before the fighting got bad – they had regularly been taken from the centres and forced to work like slaves in the houses of wealthy Libyans. He claimed female detainees were often raped and Christians targeted for particular abuse and beatings.

The UN said that it had regular access to the centres, but many detainees were never registered as refugees. They were terrified of being sold to Libyan traffickers, who often torture migrants until their families pay hefty ransoms.

Over the next few days, the clashes escalated, as militias vied for control of Tripoli. Among more than 400 migrants trapped in Ain Zara, I was told, there were eight pregnant women, and roughly twenty babies and toddlers. The whole group was abandoned by their guards, left alone to watch the city smoulder around them.

After two days without food and water, Libyan authorities sent buses to move them. I tracked their progress by GPS, as the signals from their phones edged across the map of Tripoli. At the end of their drive, I messaged the passengers that they'd reached another official detention centre.

But the change of location didn't guarantee safety. The next week, the same group was abandoned again, as their new neighbourhood became another key battleground. Over and over, I passed on their appeals to the UN, as they wondered whether they'd survive.

It's been more than a month now since that first message came through, and at least a dozen other migrants and refugees across Libya have contacted me. Some have seen friends escape detention centres, only to run into shooting in the streets. They've told me how tuberculosis kills some detainees and starvation leaves others lying motionless. They've sent me recordings of children crying and women wailing, made, they say, after bombs went off nearby. They've sent me torture videos, showing the torments of relatives held for ransom by smugglers. They say they feel abandoned by the UN, and curse the EU for 'not recognising refugees are humans too'.

I've been sent photos of a newborn baby girl, whose first breaths were taken in a vast hall surrounded by other detainees. Refugees helped her mother deliver her, wearing plastic bags on their hands instead of surgical gloves.

Through everything, the migrants carefully hide their phones, begging friends elsewhere to top up their credit, and secretly charging the batteries whenever there's electricity. 'The SIM card is our life,' one told me.

Sometimes groups crowd around a phone to craft messages together, slowly deliberating how best to describe their situation. Making people aware of their plight is the only thing giving them hope, one message explained.

A new ceasefire was announced recently. But even peace on Libya's streets may not mean any improvement for the detainees. 'We are losing hope, even in God,' the first man to contact me messaged recently. 'You can make the European people know about us,' he added, as he has many times before. 'The world should know what is happening here.'

Index